A MODERN VIEW OF
GEOMETRY

A MODERN VIEW OF GEOMETRY

Leonard M. Blumenthal
University of Missouri-Columbia

DOVER PUBLICATIONS, INC.
NEW YORK

Published in Canada by General Publishing Company, Ltd.,
30 Lesmill Road, Don Mills, Toronto, Ontario.
Published in the United Kingdom by Constable and Company, Ltd., 10 Orange Street, London WC2H 7EG.

This Dover edition, first published in 1980, is an unabridged
and corrected republication of the work originally published in
1961 by W. H. Freeman and Company.

International Standard Book Number: 0-486-63962-2
Library of Congress Catalog Card Number: 79-56332

Manufactured in the United States of America
Dover Publications, Inc.
180 Varick Street
New York, N.Y. 10014

Preface

THE TITLE of this book may recall to the reader a celebrated work of three volumes written by the German mathematician, Felix Klein, and published almost exactly fifty years ago. The second volume of Klein's *Elementare Mathematik vom Höhere Standpunkt aus* is devoted to geometry, but the "modern" viewpoint from which we shall examine certain parts of geometry has almost nothing in common with the "higher standpoint" from which that distinguished writer surveyed the subject.

The abstract, postulational, method which has now permeated nearly all parts of mathematics makes it difficult, if not meaningless, to mark the boundary of that mathematical province which is called *geometry*. There is wisdom as well as wit in saying that, geometry is the mathematics that a geometer does, for today perhaps geometry more properly describes a point of view—a way of looking at a subject—than it denotes any one part of mathematics.

Our principal concern is with the postulational geometry of planes, with the greatest emphasis on the coordinatization of affine and projective planes. It is shown in detail how the *algebraic* structure of an abstract "coordinate" set is determined by the *geometric* structure that postulates impose on an abstract "point" set. It is seen that the process can be reversed: a "geometric" entity (*plane*) arises from an initially given "algebraic" entity (*ternary ring*), and the geometric properties of the one are logical consequences of the algebraic properties of the other. The essential unity of algebra and geometry is thus made quite clear. The developments of Chapters IV, V, and VI amount to a study of ordinary analytic geometry from a "higher standpoint," and should be very illuminating to the thoughtful student.

Chapter VII contains a detailed development of a simple set of postulates for the euclidean plane, and Chapter VIII gives postulates for the non-euclidean planes.

It is believed that much of the book's content is accessible to college students and to high school teachers of mathematics, and that the first three chapters (at least) could be read with profit by the celebrated man in the street. A postulational approach to a subject usually makes very slight demands on the reader's technical knowledge, but it is likely to compensate for this by exacting from him close attention and a genuine desire to learn.

It is hoped that this book might be useful in National Science Foundation Summer Institutes. This is, indeed, an expanded version of a course given by the author at such an Institute, held at the University of Wyoming during the summer of 1959. If the book is used as a text (or for supplementary reading) in a graduate course devoted to modern postulational geometry, the first three chapters might well be omitted. On the other hand, Chapters I–V and Chapter VII constitute more than enough material for an enlightening course in a summer institute for high school teachers of mathematics.

The material of Chapters IV–VIII has been developed quite recently. Though no attempt is made to credit the source of each theorem, the author's indebtedness to R. H. Bruck's excellent expository article, "Recent Advances in the Foundations of Euclidean Plane Geometry" (American Mathematical Monthly, vol. 62, No. 7, 1955, pp. 2–24), is apparent in Chapters IV and V. Chapter VI makes considerable use of L. A. Skornyakov's monograph, "Projective Planes" (American Mathematical Society, Translation Number 99, 1953). Much of the material in both of these articles has its source in the pioneering work of Marshall Hall's "Projective Planes" (Transactions American Mathematical Society, vol. 54, 1943, pp. 229–277).

The content of Chapters VII and VIII stems from some contributions made by the author to distance geometry. A detailed account of this subject is given in his *Theory and Applications of Distance Geometry*, Clarendon, 1953.

The author wishes to record his deep appreciation of Mrs. Kay Hunt's excellent typing of the manuscript.

January 1961 Leonard M. Blumenthal

Contents

CHAPTER **I**

Historical Development of the Modern View   1

1. The Rise of Postulational Geometry. Euclid's *Elements*   1
2. Some Comments on Euclid's System   3
3. The Fifth Postulate   4
4. Saccheri's Contribution   8
5. Substitutes for the Fifth Postulate   10
6. Non-euclidean Geometry—a Nineteenth Century
   Revolution   11
7. Later Developments   14
8. The Role of Non-euclidean Geometry in the
   Development of Mathematics   18

CHAPTER **II**

Sets and Propositions   19

1. Abstract Sets   19
2. The Russell Paradox   20
3. Operations on Sets   21
4. One-to-one Correspondence. Cardinal Number   24
5. Finite and Infinite Sets. The Trichotomy Theorem and
   the Axiom of Choice   28

6. Propositions   30

7. Truth Tables   31

8. Forms of Argumentation   35

9. Deductive Theory   36

CHAPTER **III**

Postulational Systems   **38**

1. Undefined Terms and Unproved Propositions   39

2. Consistency, Independence, and Completeness of a
    Postulational System   40

3. The Postulational System 7_3 44

4. A Finite Affine Geometry   49

5. Hilbert's Postulates for Three-dimensional
    Euclidean Geometry   50

CHAPTER **IV**

Coordinates in an Affine Plane   **54**

Foreword   54

1. The Affine Plane   55

2. Parallel Classes   57

3. Coordinatizing the Plane π 58

4. Slope and Equation of a Line   60

5. The Ternary Operation T 61

6. The Planar Ternary Ring $[\Gamma, T]$ 64

7. The Affine Plane Defined by a Ternary Ring   65

8. Introduction of Addition   68

9. Introduction of Multiplication   70

10. Vectors   73

11. A Remarkable Affine Plane   74

CHAPTER **V**

**Coordinates in an Affine Plane with Desargues
and Pappus Properties** 79

Foreword (The First Desargues Property) 79

1. Completion of the Equivalence Definition for Vectors 80
2. Addition of Vectors 81
3. Linearity of the Ternary Operator 85
4. Right Distributivity of Multiplication Over Addition 88
5. Introduction of the Second Desargues Property 90
6. Introduction of the Third Desargues Property 93
7. Introduction of the Pappus Property 97
8. The Desargues Properties as Consequences of the
   Pappus Property 100
9. Analytic Affine Geometry Over a Field 104

CHAPTER **VI**

Coordinatizing Projective Planes 109

Foreword 109

1. The Postulates for a Projective Plane, and the
   Principle of Duality 110
2. Homogeneity of Projective Planes. Incidence
   Matrices of Finite Projective Planes 113
3. Introduction of Coordinates 118
4. The Ternary Operation in Γ. Addition and
   Multiplication 120
5. Configurations 121
6. Configurations of Desargues and Pappus 124
7. Veblen-Wedderburn Planes 128
8. Alternative Planes 134
9. Desarguesian and Pappian Planes 139
10. Concluding Remarks 144

CHAPTER **VII**

Metric Postulates for the Euclidean Plane   **148**

Foreword   148
1. Metric Space. Some Metric Properties of the
   Euclidean Plane   150
2. A Set of Metric Postulates. The Space $\mathfrak{M}$ 156
3. An Important Property of Space $\mathfrak{M}$ 157
4. Straight Lines of Space $\mathfrak{M}$ 160
5. Oriented Lines, Angles, and Triangles of Space $\mathfrak{M}$ 163
6. Metric Postulates for Euclidean Plane Geometry   170
7. Concluding Remarks   175

CHAPTER **VIII**

Postulates for the Non-euclidean Planes   **176**

Foreword   176
1. Poincaré's Model of the Hyperbolic Plane   177
2. Some Metric Properties of the Hyperbolic Plane   179
3. Postulates for the Hyperbolic Plane   181
4. Two-dimensional Spherical Geometry   182
5. Postulates for Two-dimensional Spherical Space   184
6. The Elliptic Plane $\mathcal{E}_{2,r}$ 185
7. Postulates for the Elliptic Plane   187

Index   189

A MODERN VIEW OF
GEOMETRY

I

Historical Development
of the Modern View

I.1. The Rise of Postulational Geometry.
Euclid's *Elements*

It is often said that geometry began in Egypt. The annual inundations of the Nile swept away landmarks and made necessary frequent surveys so that one man's land could be distinguished from that of his neighbors. These surveys did indeed give rise to a collection of geometrical formulas (many of which were merely approximations), but the Egyptian surveyors were no more geometers than Adam was a zoologist when he gave names to the beasts of the field.

For the beginnings of geometry as a deductive science we must go to ancient Greece. Through the efforts of many gifted forerunners of Euclid, such as Thales of Miletus (640–546 B.C.), Pythagoras (580?–500? B.C.), and Eudoxus (408–355 B.C.), many important geometrical discoveries were made. Plato was fond of the subject, and though he made few original contributions to it he emphasized the need for rigorous demonstrations and thus set the stage for the role Euclid was to play.

The postulational procedure that stamps so much of modern mathematics was initiated for geometry in Euclid's famous *Elements*. This epoch-making work, written between 330 and 320

B.C., has probably had more influence on the molding of our present civilization than has any other creation of the Greek intellect. Though far from attaining the perfection it aspired to, the *Elements* commanded the admiration of mankind for more than two thousand years and established a standard for rigorous demonstration that was not surpassed until modern times.

Euclid was the greatest systematizer of his age. Few, if any, of the theorems established in the *Elements* are his own discoveries. His greatness lies in his organization, into a deductive system, of all of the geometry known in his day. He sought to select a few simple geometric facts as a basis and to demonstrate all of the remainder as logical consequences of them. The facts he chose as a basis he called axioms or postulates. No proofs are given for them—they were the building blocks used to derive the theorems of his system.

We shall discuss postulational systems in Chapter II, but we may observe at once that every such system has a set of unproved statements as a basis. These statements are called *axioms* or *postulates*, or merely *assumptions*. Though Euclid applied the term axiom to so-called *common notions* that were not peculiar to geometry (for example, *if equals are added to equals, the sums are equal*) and reserved the term postulate for assumptions of a geometrical nature (for example, *two points may be joined by a line*), this distinction is not usually observed today. In this book we shall usually call our assumptions postulates.

Euclid selected five geometrical statements as the basis for his deductive treatment and introduced them in the following manner:

"Let the following be postulated:

"I. To draw a straight line from any point to any point.

"II. To produce a finite straight line continuously in a straight line.

"III. To describe a circle with any center and distance.

"IV. That all right angles are equal to one another.

"V. That, if a straight line falling on two straight lines makes the interior angles on the same side less than two right angles, the two straight lines, if produced indefinitely, meet on that side on which are the angles less than the two right angles."

I.2. Some Comments on Euclid's System

The reader will observe that Euclid's postulates involve numerous technical terms (for example, point, straight line, circle, center, right angle) as well as certain operations, such as *drawing, producing continuously,* and *falling* (a line that "falls" on two lines is called a transversal). What is the status of these notions?

Though Euclid clearly recognized the necessity for unproved propositions in his scheme, it is doubtful that he realized the necessity for *primitive* (undefined) notions. And yet it is obvious that the attempt to define everything must result either in a vicious circle or in an infinite regression. Though Euclid's thinking in this respect is not entirely clear to us, it is a fact that, in contrast with his explicitly exhibited set of postulates, his *Elements* contains no list of undefined terms, but, on the contrary, attempts to define all the terms of the subject. Some of Euclid's definitions (those for circle and right angle, for example) are precise and useful in developing the geometry, whereas others (those for point and line, for example) are vague (point is defined in terms of part, which is itself undefined) and are never used in the sequel. Perhaps they were merely intended to create images in the reader's mind. By invoking such physical notions as "drawing," "producing," and "falling," which are quite out of place in an abstract deductive system, Euclid did not attain to the modern concept of a geometry. But after all, mathematicians should have learned something in two thousand years!

A more serious criticism of Euclid's attempt to establish geometry as a deductive system is that even if he had selected certain of his notions to be primitive and had stated his five postulates without using the physical notions mentioned above, his foundation was simply insufficient to support the lofty edifice he sought to erect on it. He was able to prove many of his theorems because he used arguments that cannot be justified by his postulates. This occurs in the very first proposition, which purports to show that on any given segment A, B an equilateral triangle can be constructed. His proof is invalid (incomplete), inasmuch as it makes use of a point C whose existence is not, nor cannot be, established as a consequence of the five postulates, since the circles with centers A and B and radius AB that he employed in his argument *cannot be shown to intersect* to give the point C. This difficulty is due to the lack of a postulate that would insure the *continuity* of lines or circles.

There are other deficiencies in the *Elements*. In one of his proofs Euclid tacitly assumed that if three points are collinear, one of the points is *between* the other two, though the notion of betweenness does not appear in his basis. Euclid's concept of *congruence* of figures was the naive one of superposition, according to which he regarded one figure as being moved or placed on another. It has been conjectured that he disliked the method, but made use of it for traditional reasons or because he was unable to devise a better one.

I.3. The Fifth Postulate

The fifth (parallel) postulate of Euclid is one of the principal cornerstones on which his greatness as a mathematician rests. In his commentary on Euclid's *Elements*, Heath remarks, "When we consider the countless successive attempts made through more than twenty centuries to prove the Postulate, many of them by geometers of ability, we cannot but admire the genius of the man who concluded that such a hypothesis, which he found necessary to the validity of his whole system of geometry, was really indemonstrable."

And yet this cornerstone of Euclid's greatness was the basis of the sharpest attacks on his system. The four postulates that precede it are short, simple statements, and it is not surprising that the much more complicated nature of the assertion made by the fifth postulate suggested that it should be a theorem rather than an assumption—a view that unwittingly received some support from Euclid himself, since he did prove the postulate's converse. And so, very soon, attempts were made to remove this flaw from the *Elements*. These efforts were started in Euclid's lifetime and were continued by reputable mathematicians until the second decade of the nineteenth century, and by cranks even later. The failure of all such attempts securely established Euclid's fame and, what is more important, led to the invention of non-euclidean geometry.

All attempts to demonstrate the fifth postulate as a logical consequence of the other four postulates (amended, perhaps, to convey Euclid's meaning better than they had in their original forms) introduced, surreptitiously, assumptions that were *equivalent to the fifth postulate*, and hence, in effect, they assumed what they intended to prove. (For example, neither the infinite extent of the straight line nor the *unique* extension of segments follows from the second postulate, but Euclid likely intended that both should, since he made frequent use of those properties.) Let us examine an early and a much later example of these efforts.

Proclus (410–485 A.D.) was a very competent mathematician and philosopher who studied in Alexandria when a youth and later went to Athens, where he taught mathematics. He was highly regarded among his contemporaries for his learning and his industry. His commentary on the *Elements* is one of our chief sources of information concerning early Greek geometry, the original works of the forerunners of Euclid having disappeared. Proclus showed that the fifth postulate can be proved if it is established that (∗) *if l_1, l_2 are any two parallel lines and l_3 is any line distinct from and intersecting l_1, then l_3 intersects l_2 also.* For, assuming the italicized statement, let m_1 and m_2 denote two lines, and let m_3 denote a transversal such that the sum of the two interior angles α, β is *less* than two right angles. Then a line m_4 through P exists such that $\angle\alpha' + \angle\beta = 2$ rt $\angle$, and, on the basis of Proposition 28, Book

I of the *Elements* (*whose proof does not involve the fifth postulate*), lines m_2 and m_4 are parallel. (For all references to specific propositions of the *Elements*, see T. L. Heath, *The Thirteen Books of Euclid's Elements*, Cambridge, 1908.) Hence line m_1, which is distinct from m_4 and intersects it at P, intersects line m_2 also. Moreover, lines m_1, m_2 meet on that side of the transversal m_3 for which the sum of the interior angles α, β is less than two right angles, since if it be assumed that they meet on the other side of m_3, they would form with m_3 a triangle with an exterior angle α that is *less* than an opposite interior angle β', which is contrary to Proposition 16, Book I (whose proof does not involve the fifth postulate).

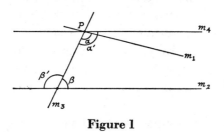

Figure 1

This much having been done in an acceptable way, it remained for Proclus to derive the above proposition (∗) from Postulates I–IV. Here is his argument, as translated by Heath.

"Let AB, CD be parallel, and let EFG cut AB; I say that it will cut CD also.

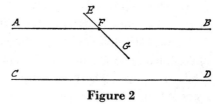

Figure 2

"For, since BF, FG are two straight lines from one point F, they have, when produced indefinitely, a distance greater than any magnitude, so that it will also be greater than the interval between the parallels.

"Whenever, therefore, they are at a distance from one another greater than the distance between the parallels, FG will cut CD."

The assertion made in the second sentence of Proclus' "proof" is essentially an axiom he ascribed to Aristotle. It was *not* a postulate of the *Elements*, nor did Proclus attempt to derive it from

Postulates I–IV. But a deeper criticism of Proclus' argument concerns the tacit assumption embodied in the phrases "the interval between the parallels" and "the distance between the parallels." Those phrases imply that parallel lines are at a constant distance from one another, but the justification of this implicit assumption is Postulate V itself, which is, indeed, its logical equivalent! For an argument based on Aristotle's axiom it would suffice that the perpendicular distances from points on AB to CD be merely bounded, but this also is equivalent to Postulate V.

Bernhard Friedrich Thibaut (1775–1832) employed a method of "demonstrating" the fifth postulate (published in his book, *Grundriss der reinen Mathematik*, 3 ed., Göttingen, 1818) that turned away from properties of parallels (which had been featured in many earlier attempts) and proceeded in a novel manner.

It is shown in the *Elements* (using the fifth postulate) that the sum of the angles of any triangle equals two right angles, and it was known in Thibaut's day that, conversely, the fifth postulate could be proved if it was assumed that a single triangle exists with angle-sum equal to two right angles. This had been established by the French mathematician, A. M. Legendre, prior to 1823. Thibaut's approach to the fifth postulate was by way of Legendre's angle-sum theorem, and he reasoned as follows:

Rotate the triangle ABC about vertex A through the angle CAC', then rotate triangle $AB'C'$ about point B through the angle ABA'', and finally rotate triangle $A''B''C''$ about point C through angle BCA'''.

As a result of these three rotations, triangle ABC has been turned through four right angles, and hence, according to Thibaut,

(†) $$\angle CAC' + \angle ABA'' + \angle BCA''' = 4 \text{ rt } \angle.$$

Since each of the three angles in the left member of this equality is an exterior angle of triangle ABC and is, consequently, equal to two right angles minus the corresponding interior angle of that triangle, substitution in (†) gives the angle-sum of triangle ABC equal to two right angles.

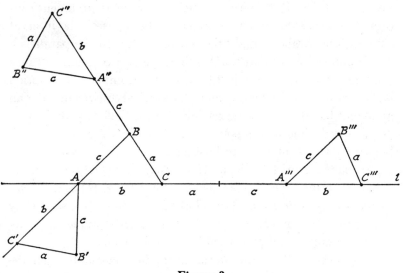

Figure 3

If, as a result of the three rotations, triangle ABC had been rotated about a single point into itself, Thibaut's demonstration
would be acceptable. But the triangle has been *translated* along
line l (through the distance $a + b + c$) as well as rotated through
four right angles and the argument takes no account of this translation. There is no justification for this in Postulates I–IV, and,
indeed, it may be shown that assuming every motion may be resolved into a rotation and a translation *independent of it* is equivalent to assuming Postulate V itself! Thibaut's procedure can be
carried out for spherical triangles, and in that case it is clear that
the translation *must* be taken into account, since it is, in fact, a
rotation of the triangle about the pole of one of its sides.

I.4. Saccheri's Contribution

The most elaborate attempt to prove the fifth postulate, and
the most far-reaching in its consequences, was made by the Italian
priest, Girolamo Saccheri (1667–1733), who taught mathematics at

the University of Pavia. His great work, *Euclides ab omni naevo vindicatus sive conatus geometricus quo stabiliuntur prima ipsa geometriae principia*, was published in 1733. It apparently failed to attract much attention, for it was soon forgotten, and there is no evidence that Gauss, Bolyai, or Lobachewsky—the founders of non-euclidean geometry—ever heard of the book or its author. But Saccheri's work establishes him as an important contributor to the development of that subject.

As stated in the title of the book, Saccheri's aim was to free Euclid from all error—in particular (and most important), from the error of having assumed the fifth postulate. His novel procedure for accomplishing this introduced an important figure into geometry: the Saccheri quadrilateral. Saccheri erected, at the endpoints A, B of a segment, equal segments AC and BD, each perpendicular to segment AB, and joined points C, D by a straight line. On the basis of Postulates I–IV it is easily proved that $\angle ACD = \angle BDC$, for if P,

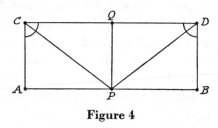

Figure 4

Q denote the midpoints of segments AB, CD, respectively, the two right triangles ACP and BDP are congruent (Proposition 4, Book I), so $\angle ACP = \angle BDP$, and side PC = side PD. Then the sides of triangle CPQ are equal, respectively, to the sides of triangle DPQ, and, consequently, these two triangles are congruent (Propositions 4, 8, Book I). (These propositions are proved without the aid of the fifth postulate, which was invoked by Euclid for the first time in the proof of Proposition 29, Book I.) It follows that $\angle PCD = \angle PDC$, and, consequently,

$$\angle ACD = \angle ACP + \angle PCD = \angle BDP + \angle PDC = \angle BDC.$$

Calling the equal angles at C and D the *summit angles* of the Saccheri quadrilateral, the following three possibilities are exhaustive and pairwise mutually exclusive: (1) *the summit angles are right angles*, (2) *they are obtuse angles*; (3) *they are acute angles.* Saccheri called these the right-angle hypothesis, the obtuse-angle hypothesis, and the acute-angle hypothesis, respectively, and he proved that if any one of these hypotheses were valid for one of his quadrilaterals, it was valid for every such quadrilateral. Using the infinitude (unbounded length) of the straight line, he showed that the fifth postulate is a consequence of the right-angle hypothesis and that the obtuse-angle hypothesis is self-contradictory. There remained only to dispose of the acute-angle hypothesis!

Saccheri explored the consequences of the acute-angle hypothesis, hoping to reach a contradiction. But though he obtained many results that seemed strange (because they differed markedly from those that had been established by use of the fifth postulate), he never succeeded in his search for the desired contradiction. Unable to cast out the acute-angle hypothesis on purely logical grounds, and so prove the fifth postulate, Saccheri took refuge on the weaker terrain of intuition and concluded, in proposition XXXIII of his book, that "the hypothesis of the acute angle is absolutely false, because it is repugnant to the nature of a straight line."

It is doubtful that this vague, inglorious conclusion of an otherwise clear and logical investigation satisfied Saccheri. We know now that, contrary to Saccheri's fixed idea, no logical contradiction can be deduced from the acute-angle hypothesis, for it gives rise to a geometry that is very different from that of Euclid, but just as consistent. In uncovering the consequences of the acute-angle hypothesis, the Italian priest was unknowingly developing a new geometry. Instead of finding, as he thought, a route to the proof of the fifth postulate (his avowed objective) he was actually discovering a new world!

I.5. Substitutes for the Fifth Postulate

The following are some of the best-known alternatives for the fifth postulate. Some of these statements were explicitly proved by their authors to be equivalent to Euclid's postulate, whereas others were implicit, tacit, assumptions made when demonstrations of the postulate were attempted.

Proclus. Parallels remain at a finite distance from one another.

The most frequently used alternative appeared in 1795 in a treatise on the first six books of Euclid by the Englishman, John Playfair. The popularity of that work—ten editions were published, the last one in 1846—attached Playfair's name to the statement (though it is hardly more than a paraphrase of one attributed to Proclus) and perhaps accounts for its favor today.

Playfair. Through a given point, not on a given line, only one parallel can be drawn to the given line.

Though this statement seems simpler than the fifth postulate, there are several reasons why Euclid's assertion should be preferred. It suffices here to point out that the fifth postulate gives at once a direct method of determining when two lines intersect—an extremely important matter in the development of any geometry.

Legendre. There exists a triangle in which the sum of the three angles is two right angles.

Laplace, Saccheri. There exist two non-congruent triangles with the angles of one equal, respectively, to the angles of the other.

Legendre, Lorenz. Through any point within an angle less than two-thirds of a right angle, there is a line that meets both sides of the angle.

Gauss. If K denotes any integer, there exists a triangle whose area exceeds K.

Bolyai. Given any three points not on a straight line, there is a circle that passes through them.

I.6. Non-euclidean Geometry—a Nineteenth Century Intellectual Revolution

The efforts of two thousand years to change the status of Euclid's famous assertion from a postulate to a theorem resulted in failure with regard to that objective, but they were highly successful in another respect. They served to re-orient men's thinking concerning the nature of geometry, and they were the necessary preparation for a cultural environment that would enable its beneficiaries to accomplish work of far greater importance.

Perhaps the first individual to have a clear conception of a geometry other than Euclid's—one in which the fifth postulate is actually denied—was Karl Friedrich Gauss (1777–1855), of Göttingen, Germany, the greatest mathematician of the nineteenth century, if not of all time. While still in his twenties, Gauss began his study of the theory of parallels—a study he carried on for more than thirty years. He soon understood the profound nature of the difficulties that prevented him from demonstrating the fifth postulate, and after considerable thought he formulated a new geometry, which he called *non-euclidean*, and started its development. In a letter to his friend, Franz Adolf Taurinus, dated November 8, 1824, he states: "The assumption that the angle-sum [of a triangle] is less than 180° leads to a curious geometry, quite different from ours [the euclidean] but thoroughly consistent, which I have developed to my entire satisfaction. The theorems of this geometry appear to be paradoxical, and, to the uninitiated, absurd, but calm, steady reflection reveals that they contain nothing at all impossible."

But Gauss never published any of his great discoveries in this field. In the letter to Taurinus, he cautions his friend to "consider it a private communication of which no public use, or use leading in any way to publicity, is to be made." Not until seven years later, in 1831, did Gauss write a short account of the new geometry, which was found among his papers after his death. Perhaps a communication he received on February 14, 1832, from Wolfgang Bolyai, the Hungarian geometer, was the reason he did not prolong the report of his early investigations.

Before turning our attention to that communication let us observe that we are not in a position today to understand the difficulties that would have beset even a man of Gauss' stature, had he published his non-euclidean geometry when he first formulated it. The traditionalism and authoritarianism that held all independent thinking in bondage during the Middle Ages had by no means been entirely overthrown. The immense authority of the German philosopher, Immanuel Kant, who died in 1804, supported the doctrine that euclidean geometry was *inherent in nature*. (Thus, though Plato had said merely that God geometrizes, Kant asserted, in effect, that *God geometrizes according to Euclid's Elements*.) Gauss shrank from the controversy in which the publication of his new geometry would have involved him. He had more important ways to spend his time.

Let us now return to the communication from Wolfgang Bolyai that Gauss received in February, 1832. They had been fellow students of the University in Göttingen and had doubtless often discussed those fascinating problems arising from the fifth postulate. After Bolyai left Göttingen (Gauss spent his whole life there) the two friends exchanged occasional letters, in one of which Gauss pointed out an error that invalidated one of Bolyai's "proofs" of the postulate.

The chief mathematical work of Wolfgang Bolyai was a two-volume treatise on geometry, which he published in 1832–1833, but his greatest contribution to mathematics was his son Johann. The young Bolyai wrote a 26-page appendix for his father's treatise, in which he gave an account of investigations he had begun nearly ten years before, when he was only 21 years old. It was an advance copy of this appendix that Gauss received in February, 1832, and which caused him to abandon his project of writing his own results "so that they would not perish" with him.

Using Playfair's form of the fifth postulate, Johann Bolyai investigated the consequences of denying it by assuming that either no parallel or more than one parallel existed. The first alternative was easily rejected (just as Saccheri's obtuse-angle hypothesis had been); it was the second alternative that led to the interesting new geometry (the geometry of Saccheri's acute-angle hypothesis). But the points of view of Bolyai and Saccheri were quite different. Whereas the latter believed a clear-cut contradiction would be found if his investigation of the acute-angle hypothesis were only pushed far enough, the former was convinced that in developing the assumption of more than one parallel he was founding a new kind of geometry.

Gauss' reception of the appendix was very disappointing to the young man. Though Gauss wrote to another friend, C. L. Gerling, "I consider the young geometer Bolyai a genius of the first rank" (high enough praise for any man, it would seem, coming from Gauss), Bolyai was not happy to read in a letter to his father, "the entire content of the work, the path which your son has taken, the results to which he is led, coincide almost exactly with my own meditations which have occupied my mind for from thirty to thirty-five years." So Gauss did it all thirty years ago, Johann might have sadly mused, there is nothing new under the sun.

Nor was this the end of Bolyai's disappointments, for in 1848 he learned that a professor at the University of Kasan, Nikolai Ivanovich Lobachewsky (1793–1856), had also discovered the new geometry and had even obtained priority by publishing his results in 1829.

Thus, when the culture was ripe for it, three men, Gauss, Bolyai, and Lobachewsky (a German, a Hungarian, and a Russian), arose in widely separated parts of the learned world, and, working independently of one another, created a new geometry. It would be difficult to overestimate the importance of their work. They broke the chains of Euclidean bondage; their invention has been hailed as "primate among the emancipators of the human intellect" and as "the most suggestive and notable achievement of the last century." A significant milestone in the intellectual progress of mankind had been passed.

I.7. Later Developments

The first phase in the development of non-euclidean geometry culminated a little more than one hundred years ago with the pioneering work of Gauss, Bolyai, and Lobachewsky. A feature of the second phase was a paper published in 1868 by the Italian geometer, Eugenio Beltrami (1835–1900), which gave a final answer to the question of the consistency (freedom from contradiction) of the new geometry. The reader will recall that Saccheri was convinced that the acute-angle hypothesis, which gave rise to the geometry, would lead to a contradiction, and when he was unable to establish this he rejected the hypothesis on aesthetic rather than logical grounds as being "repugnant to the nature of a straight line." The perplexing question had not been answered by any of the great trio of founders. It seems that Bolyai harbored the suspicion that extending his investigations to three-dimensional space might reveal inconsistencies, and that Lobachewsky entertained similar fears concerning his own development. It is a commentary on the state of mathematics of that day to observe that the question of the consistency of euclidean geometry itself apparently did not arise. Only geometries that differed from the euclidean were suspect and had to be proved innocent of the crime of inconsistency. But the day was approaching when even euclidean geometry had to meet that challenge.

Beltrami's paper gave an interpretation of non-euclidean geometry as being the geometry on a certain class of surfaces in euclidean three-space. Hence the paradoxical properties of the new geometry are actually realized on these surfaces, and so *any inconsistency in the new geometry is also an inconsistency in euclidean geometry.* The new geometry is thus as consistent as the old one, and Euclid was at long last vindicated from all error.

We have observed that the term non-euclidean geometry was applied by Gauss to that geometry obtained when Euclid's fifth postulate is replaced by its *negation* and all of the remaining postulates (the four that Euclid stated explicitly, and others that he tacitly introduced) are unaltered. We have noted that the assumption of no parallel (the obtuse-angle hypothesis of Saccheri) contradicts the infinite extent of the straight line, which Euclid made use of, and so the multi-parallel case (acute-angle hypothesis) is the only alternative to euclidean geometry.

If, however, the infinite extent of the straight line is *not* assumed, then another geometry arises in which no parallels exist. It is customary today to extend the term non-euclidean to include this geometry also. The German mathematician, Felix Klein (1849–1925), attached to these geometries the names currently used for them. The original non-euclidean geometry of Saccheri, Gauss, Bolyai, and Lobachewsky he called *hyperbolic geometry*, the geometry with no parallels (the obtuse-angle hypothesis) *elliptic geometry*, and euclidean geometry *parabolic geometry*. This terminology arose from the projective approach to the non-euclidean geometries due to Klein and the British mathematician, Arthur Cayley (1821–1895), and was motivated by the fact that the number of infinite points on a straight line is two, none, or one, according to whether the acute-angle, obtuse-angle, or right-angle hypothesis, respectively, holds.

I.8. The Role of Non-euclidean Geometry in the Development of Mathematics

In the last paragraph of Section I.6 it is stated that the invention of non-euclidean geometry was "the most suggestive and notable achievement of the last century." This judgment was expressed by the great German mathematician, David Hilbert (1862–1943). What justifies such an assertion?

As late as the second decade of the nineteenth century geometry was thought of as an idealized description of the spatial relations of the world in which we live. In writing his *Elements*, Euclid chose for his postulates statements that had their roots in common experience. He thought of them as being true, self-evident (dictionaries still give that now-obsolete meaning for "axiom"), and factual. Euclid's task was to proceed from those simple facts and establish more complicated ones which, like the postulates, would be regarded as idealizations of the way in which the physical world behaved. It follows as a corollary of viewing geometry in this light that its assumptions have an *a priori* character and that any geometrical statement is either *true* (nature behaves that way) or *false* (nature does not behave that way). Thus the older concept of geometry makes it a branch of physics rather than a part of mathematics (in the modern sense of that word).

If a postulate is regarded as a true statement, its negation is evidently false, and so, if the negation is assumed true, a contradiction should result. Doubtless, this is what convinced Saccheri, for example, that the acute-angle hypothesis must lead to a contradiction if only its consequences were developed far enough. But when Beltrami proved that no contradiction is introduced by the acute-angle hypothesis, and, consequently, that there exists another geometry, as consistent as Euclid's, whose theorems are often in disagreement with the corresponding ones in the *Elements*, then the view of geometry as a description of actual space must be abandoned!

Perhaps the first consequence of abandoning the classical viewpoint is that the postulates of geometry lose their apodictic character—they no longer express necessary truths. But a more fundamental change wrought by the existence of new geometries affected the basic notion of truth itself. Is it "true" that the sum of the angles of every triangle equals two right angles? In euclidean geometry it is proved that the angle-sum is, indeed, equal to two right angles, but in hyperbolic geometry the angle-sum is always *less* than two right angles, and in elliptic geometry the angle-sum is always *greater* than two right angles. In still another geometry it is not possible to prove that the angle-sum of every triangle equals two right angles, and *it is equally impossible to prove the negation* (that is, that there is at least one triangle whose angle-sum is *not* equal to two right angles). What, then, does "truth" mean in mathematics?

Thus the consequences of abandoning the view that geometry is a description of actual space, made necessary by the existence of several equally good geometries, were very far-reaching. They made it necessary to give new answers to such questions as, What is a geometry?, What is the nature of the postulates on which a geometry rests?, and even, What is the nature of *truth* in mathematics? We shall consider these matters in the next chapter, which is devoted to a discussion of the modern view of a postulational system.

The great strides made by mathematics during the past sixty years are due in large measure to the introduction of the postulational method in many parts of the subject. The use of this method in algebra and analysis is, of course, a direct consequence of its successful application to geometry—a success made possible by the labors of Euclid, by the two thousand years of effort devoted to proving the fifth postulate, and by the supreme achievements of the founders of the non-euclidean geometries.

II

Sets and Propositions

II.1. Abstract Sets

The notion of a set is deceptively simple. It is simple because nearly everyone has an understanding of the meaning of the term (or of one of its synonyms: collection, class, aggregate)—an understanding that is, moreover, precise enough for ordinary purposes. However, its simplicity is deceptive because close examination of the notion has revealed that unless the concept is suitably restricted it gives rise to contradictions and is therefore not a fit object for mathematical study.

Perhaps most people would agree that all persons living at this moment, who were born in the State of Missouri, form a set. But suppose the birth certificate of one of our friends, Mr. Smith, has been lost, and there is no way of ascertaining whether or not Mr. Smith was born in Missouri. Then there is no possibility of deciding whether or not Mr. Smith is a member of that set. The impossibility of resolving the issue of membership, in the particular case of Mr. Smith, is not crucial. What is decisive is that a criterion has been laid down (birth in Missouri) that every individual of a certain universe (all persons living at this moment) does or does not satisfy. Such a dichotomous criterion or law defines a set; indeed, some writers identify a set with the law that defines its elements, much as a (real) function of a real variable is identified with a law that pairs one real number with another.

We shall soon give an example of one kind of difficulty that a completely unrestricted set concept evokes, but we shall not enter into the logical difficulties of the notion. It suffices for our purpose to demand merely that the statement "R denotes a set" implies: (1) there is no alternative for an entity other than to belong or not to belong to R, and (2) if two elements of R are denoted by a, b, there is no alternative other than $a = b$ (that is, a and b are notations for the same element of R) or $a \neq b$ (that is, the elements denoted by a, b are not the same).

Most of the operations on sets used in mathematics are quite independent of the nature of the elements of the sets. An *abstract set* is simply a set, the nature of whose elements is unspecified or is voluntarily ignored.

II.2. The Russell Paradox

Perhaps the most readily understood contradiction that is encountered by permitting an intuitive concept of set is the following one, devised by the British logician, Bertrand Russell (1872–   ). It is likely that every set familiar to the reader has the property that it is *not* a member of itself (for example, the set of all elephants is not an elephant, the set of all whole numbers from 1 to 10 is not a whole number from 1 to 10, and so on). Let us call a set *ordinary*, provided it is *not* a member of itself.

Are all sets ordinary? It seems not, for the set of all ideas is itself an idea, and hence contains itself as a member, and the set composed of everything not a man is itself not a man, and so contains itself as a member. Perhaps the reader would like to think of additional examples of such sets. We will call a set *extraordinary*, provided it *is* a member of itself.

Now it would seem that every set is either an ordinary set or an extraordinary set and that no set is both ordinary and extraordinary. But let us examine the *set of all ordinary sets*. If this set is an ordinary set, it is a member of itself, and, consequently, it is an extraordinary set, and if this set is an extraordinary set, it is a member of itself, and, consequently, it is an ordinary set! Thus if the set is *either* ordinary or extraordinary, it is *both* ordinary and extraordinary.

The following is an amusing variant of the Russell paradox. Before card-index files came into use, each library contained a volume in which the title of every book in the library was listed. This volume was known as the catalogue. Since the catalogue was one of the library's books, doubtless some librarians listed *it* in the catalogue, whereas others did not. With the encouragement of a large grant from a Foundation (to bring the story up-to-date) someone set about making a catalogue of all (and only) catalogues *that did not list themselves.* The work progressed very well until the fateful day when the cataloguer had to decide whether the catalogue he was compiling should list itself. If that catalogue did *not* list itself, then it *must* list itself since its purpose was to list *every* catalogue that did not list itself. On the other hand, the new catalogue must *not* list itself, since its purpose was to list *only* those catalogues that did not list themselves!

Contradictions in set theory are often caused by assuming a set may contain elements whose definitions require the existence of the set as a completed whole. Thus the "set" of all propositions might be supposed to contain as an element the proposition, "All propositions are false." But this element obviously requires that the set of all propositions be a completed entity, which it certainly is not if "new" propositions about it may be created at will.

A principle used to avoid illegitimate sets is, *Whatever involves all of a set must not be one of its elements.* This principle may be invoked to invalidate such "sets" as: the set of all sets; the set of all propositions; the set of all things I like.

II.3. Operations on Sets

To express that an entity, denoted by x, is a member of a set, denoted by A (that is, A contains x as an element), we write $x \in A$, which is read, "x is an element of A." If A, B denote sets, and

$x \in A$ implies $x \in B$, we say A is a *subset* of B, and we write $A \subset B$. Note that $A \subset B$ does not exclude the possibility $B \subset A$. If $A \subset B$ and $B \subset A$, clearly the sets A, B consist of the same elements, and we write $A = B$. If $A \subset B$ and $A \neq B$, we say A is a *proper* subset of B. Thus the set of even integers is a proper subset of the set of all integers.

If A, B, C denote sets and $x \in C$ if and only if $x \in A$ or $x \in B$, C is called the *sum (union)* of A, B, and we write $C = A \cup B$. Note that "or" is used here in the *inclusive* sense; that is, $x \in A$ or $x \in B$ is true of any entity x that is a member of both sets A, B. It follows from the definition of set sum that

(1) $A \cup B = B \cup A$;

(2) $A \cup (B \cup C) = (A \cup B) \cup C$;

(3) $A \cup A = A$;

that is, addition of sets is a *commutative, associative,* and *idempotent* operation.

The set of all black objects may be added to the set of all cats to produce the set of all objects, each of which is either black or a cat (or both).

If A, B, C are sets and $x \in C$ if and only if $x \in A$ *and* $x \in B$, C is called the *product* of A, B, and we write $C = A \cap B$. If A is the set of all black objects and B is the set of all cats, the set $A \cap B$ consists of all black cats.

We may always form the sum of two sets, but if sets A and B have no common members (as, for example, the set of all even integers and the set of all odd integers), is their product defined? The reader will recall that to each set there corresponds a law or rule or qualification that determines whether or not a given entity is a member of the set. But we have seen that an unrestricted notion of set leads to contradictions, so it is necessary to place certain restrictions on a law or a rule in order that the entities conforming to it shall properly be said to constitute a set. Now a qualification such as "all living centaurs" is an admissible one. The set of all entities conforming to it *has no members*, and it is called a *null* set. It follows from the definition of equality for sets, given above, that each two null sets are equal, and, consequently, we speak of *the* null set. Hence each two sets have a (unique) product, for if they have no members in common, their product is the null set. It is easily shown that set multiplication is *commutative, associative,* and *idempotent;* that is,

(1) $A \cap B = B \cap A;$

(2) $(A \cap B) \cap C = A \cap (B \cap C);$

(3) $A \cap A = A.$

The operations of sum and product are connected by two distributive relations. Not only does multiplication distribute over addition (as it does for real numbers) but addition distributes over multiplication (as it does *not* for real numbers). That is, for any three sets $A, B, C,$

$$(*) \qquad A \cap (B \cup C) = (A \cap B) \cup (A \cap C),$$

$$(**) \qquad A \cup (B \cap C) = (A \cup B) \cap (A \cup C).$$

We give the proof of $(**)$ and leave $(*)$ to be proved as an exercise. If $x \in A \cup (B \cap C),$ then $x \in A$ or $x \in B \cap C.$ The first alternative yields $x \in A \cup B$ *and* $x \in A \cup C,$ so $x \in (A \cup B) \cap (A \cup C);$ the second alternative gives $x \in B$ *and* $x \in C,$ so $x \in A \cup B$ and $x \in A \cup C.$ Hence from $x \in A \cup (B \cap C)$ follows $x \in (A \cup B) \cap (A \cup C),$ and, consequently, $A \cup (B \cap C) \subset (A \cup B) \cap (A \cup C).$

Now suppose $x \in (A \cup B) \cap (A \cup C).$ Then $x \in A \cup B$ and $x \in A \cup C;$ that is, $x \in A$ or $x \in B$ and $x \in A$ or $x \in C.$ From the four possibilities presented here we conclude $x \in A$ or $x \in B \cap C,$ from which $x \in A \cup (B \cap C).$ Hence $(A \cup B) \cap (A \cup C) \subset A \cup (B \cap C),$ which, taken together with the last relation of the preceding paragraph, gives $A \cup (B \cap C) = (A \cup B) \cap (A \cup C).$

Let S denote a given set, and let S^* denote the set of all subsets of $S;$ that is, each *element* of S^* is a *subset* of $S,$ and each subset of S is an element of $S^*.$ For example, if S consists of exactly three elements a, b, c (we write $S = \{a, b, c\}$), then $S^* = \{\{a\}, \{b\}, \{c\}, \{a, b\}, \{a, c\}, \{b, c\}, \{a, b, c\}, \Omega\},$ where Ω denotes the null set and

$\{a\}$ denotes the set with a as its only element, and so on. Note that we distinguish between the element a of S and the set $\{a\}$, which is an element of S^*, for suppose each set consisting of just one element could be identified with that element. If E denotes an arbitrary set, E could be identified with the set $\{E\}$ whose only element is E, and from $E \in \{E\}$ would follow $E \in E$; that is, each set E would be an element of itself!

If $A \in S^*$, then S^* contains as an element that subset of S which consists of all elements of S that are *not* elements of A. Let A' denote that element of S^*. It is called the *complement* of A (in S), and clearly for every element A of S^*: $A \cup A' = S$, and $A \cap A' = \Omega$.

The following very useful relations are known as the DeMorgan formulas [after the British logician and mathematician, Augustus DeMorgan (1806–1871)].

(†) $(A \cup B)' = A' \cap B',$
(††) $(A \cap B)' = A' \cup B'.$

We prove (†) and leave the proof of (††) as an exercise. If $x \in (A \cup B)'$, then $x \in S$ and $x \notin A \cup B$ ($\notin$ is read "is not an element of"). From $x \notin A \cup B$ follows $x \notin A$ *and* $x \notin B$, and so $x \in A'$ and $x \in B'$. Consequently, $(A \cup B)' \subset A' \cap B'$, and, similarly, $A' \cap B' \subset (A \cup B)'$.

● EXERCISES

1. If A, B, C denote sets, show that (i) $A \cap (B \cup C) = (A \cap B) \cup (A \cap C)$ and (ii) $(A \cap B)' = A' \cup B'$.

2. If a set S consists of exactly n elements, show that the set S^*, of all subsets of S, has exactly 2^n elements.

3. Show that $A \cap B = A$ if and only if $A \cup B = B$, and that each relation is equivalent to $A \subset B$.

II.4. One-to-one Correspondence. Cardinal Number

One of the most important notions of modern mathematics is that of one-to-one correspondence—a relation that exists between the elements of one set S and the elements of another set T, provided that to each element s of S there is associated exactly one element t of T and that each element of T is the associate of exactly one element of S. Such a correspondence effects a *pairing* of the elements of S with the elements of T. For example, if each seat in an auditorium is occupied by a spectator and no spectator is standing, a one-to-one correspondence between the elements of the set of seats in the auditorium and the elements of the set of spectators is established by associating with each seat the spectator occupying it. No such correspondence exists in the event that (a) there are empty seats and no spectator standing, or (b) there are spectators standing and no empty seat.

The German mathematician, Georg Cantor (1845–1918), based his theory of cardinal numbers, finite and infinite, on the notion of one-to-one correspondence. He defined two sets to have the *same* cardinal number, provided there exists a one-to-one correspondence between the elements of one set and the elements of the other. Some surprising results are obtained when this definition is applied to certain sets. For example, the set N of *all* natural numbers 1, 2, 3, $\cdots$, n, $\cdots$ has the same cardinal number as the set N_e of all *even* natural numbers 2, 4, 6, $\cdots$, $2n$, $\cdots$, since a one-to-one correspondence between their elements is established by associating with each number of N its double, which is a number of N_e. The set N_e is clearly a proper subset of the set N, nevertheless, the two sets have the same cardinal number! So Euclid's axiom, "The whole is greater than any of its parts," is not always valid, even under reasonable interpretations of "greater."

Recalling that a rational number is the quotient of two integers, it is easy to prove the even more surprising result that *the set of all rational numbers has the same cardinal number as the set of all natural numbers* (even though there are infinitely many rational numbers between any two natural numbers).

Each rational number has a *unique* representation as an irreducible fraction p/q, where p denotes an integer (positive, negative, or zero) and q denotes a natural (whole) number. We shall agree to use 0/1 for the unique representation of 0.

Now we may classify the rational numbers by putting in the k-th class all *irreducible* fractions p/q for which

$$|p| + q = k,$$

where $|p|$ denotes the absolute value of the integer p. Clearly, every rational number belongs to exactly one of these classes. Each of these classes contains just a finite number of rationals, since, for example, all of the members of the k-th class may be obtained from the $2k - 1$ numbers

$$-(k - 1)/1, -(k - 2)/2, \cdots, -1/(k - 1), 0/k, 1/(k - 1),$$
$$2/(k - 2), \cdots, (k - 1)/1$$

by deleting those fractions that are not irreducible.

We order the numbers in each class according to magnitude and write in a line all numbers of the first class, followed by all numbers of the second class, followed by all numbers of the third class, and so on, to obtain the sequence

$$0/1, -1/1, 1/1, -2/1, -1/2, 1/2, 2/1, -3/1, -1/3,$$
$$1/3, 3/1, \cdots,$$
$$1, \quad 2, \quad 3, \quad 4, \quad 5, \quad 6, \quad 7, \quad 8, \quad 9,$$
$$10, \quad 11, \cdots,$$

in which every rational number occurs just once. Finally, by pairing each number in the sequence with the symbol of the ordinal number that denotes its place in the sequence, as shown above, we obtain a one-to-one correspondence between the set of all rational numbers and the set of all natural numbers.

This result is a shock to our intuition and might lead us to conjecture that every infinite set has the same cardinal number as the set N of all natural numbers. But such a conjecture would be false, for we will now show that *there does not exist any one-to-one correspondence between the set N and the set of all real numbers*.

Let us suppose such a correspondence does exist, and let us denote the real number r that corresponds to the natural number n

by r_n. Then every real number occurs exactly once in the sequence $r_1, r_2, \cdots, r_n, \cdots$.

Clearly, every real number r can be written in *one and only one* way as the sum of the greatest integer (positive, negative, or zero) that is not greater than r [this integer being denoted by $I(r)$] and a *non-negative* infinite decimal fraction in which *not* all but a finite number of digits are 9. For example, $-2.345 = -3 +$ $.655000 \cdots$, $.213 = .213000 \cdots$, $.12999 \cdots = .13000 \cdots$. Writing the members of our sequence in that way,

$$r_n = I(r_n) + 0.c_1^{(n)}c_2^{(n)}c_3^{(n)} \cdots, \quad n = 1, 2, 3, \cdots,$$

we consider the infinite decimal fraction $0.c_1c_2c_3 \cdots$ whose digits are defined as follows:

$$c_n = 0, \text{ if } c_n^{(n)} \neq 0,$$
$$c_n = 1, \text{ if } c_n^{(n)} = 0, n = 1, 2, 3, \cdots.$$

It is clear that $0 + 0.c_1c_2c_3 \cdots$ is the kind of representation for a real number used above (indeed, in the non-negative infinite decimal $0.c_1c_2c_3 \cdots$ *no* digit is 9), but it is different from every one of the numbers $r_n = I(r_n) + 0.c_1^{(n)}c_2^{(n)}c_3^{(n)} \cdots$ of the sequence $r_1, r_2, \cdots, r_n, \cdots$, since, for every $n = 1, 2, 3, \cdots$, its n-th digit c_n differs from the n-th digit $c_n^{(n)}$ of the decimal part of r_n.

Hence the sequence $r_1, r_2, \cdots, r_n, \cdots$ does not contain every real number, and so there does not exist a one-to-one correspondence between the set N of natural numbers and the set R of all real numbers. Hence N and R do not have the same cardinal number.

So far we have managed with the notion of *same* cardinal number without actually defining the cardinal number of a set. Let us suppose all sets are classified by putting in each class all sets with the same cardinal number. Then each set M belongs to exactly one of these classes, and the *cardinal number* of M, denoted by $\overline{\overline{M}}$, is that class. Thus the class of all pairs is the cardinal number denoted by 2. The symbol $\aleph_0$ (read "aleph nought") is customarily used to denote the cardinal number of the set of all natural numbers, and c denotes the cardinal number of the set of all real numbers. We have shown that $\aleph \neq \mathsf{c}$.

● EXERCISE

Establish geometrically a one-to-one correspondence between the points of a segment one unit long and the points of a segment two units long.

II.5. Finite and Infinite Sets.   The Trichotomy Theorem and the Axiom of Choice

The terms finite and infinite have, in the foregoing discussion, been applied to sets, and it has been left to the reader to interpret these terms in the light of his knowledge or intuition. We may define a set F to be *finite*, provided either it is null or a natural number n exists such that F has the same cardinal number as the set $\{1, 2, 3, \cdots, n\}$, and then call a set *infinite*, provided it is *not* finite. These are, perhaps, the most natural definitions for those concepts, but there is another interesting way to proceed.

We have noted that the set N has the same cardinal number as one of its proper subsets, the set N_e. The German mathematician, Richard Dedekind (1831–1916), took this property to be characteristic of infinite sets; that is, a non-null set is infinite, provided there exists a one-to-one correspondence between its elements and those of a proper subset. All other sets he called finite. It can be shown that the two notions are equivalent: a set is finite or infinite, in the sense of Dedekind, if and only if it is finite or infinite, respectively, according to the first definition.

A set whose cardinal number is $\aleph_0$ is called *denumerable*. In addition to the set of natural numbers and the set of rational numbers, other important denumerable sets are: (1) the set of algebraic numbers (these are numbers, real or imaginary, that are roots of algebraic equations $a_0 x^n + a_1 x^{n-1} + \cdots + a_{n-1} x + a_n = 0$, where n is any natural number and $a_0, a_1, \cdots, a_n$ are integers); and (2) the set of all rational points of n-dimensional space, n being a natural number (these are points with all coordinates rational).

Cardinal numbers of infinite sets are called *transfinite* cardinals. The cardinal $\aleph_0$ is the first transfinite cardinal encountered. Before we may say $\aleph_0$ is the *smallest* transfinite cardinal we must define a means of comparing such cardinals.

If $\mathfrak{m}$, $\mathfrak{n}$ denote any two cardinals (finite or transfinite), let M, N be any two sets such that $\overline{\overline{M}} = \mathfrak{m}$ and $\overline{\overline{N}} = \mathfrak{n}$. We define $\mathfrak{m} < \mathfrak{n}$, provided there exists a one-to-one correspondence of M with a subset of N, but there does *not* exist a one-to-one correspondence of N with a subset of M. It follows at once that any two cardinals satisfy *at most one* of the relations

$$\mathfrak{m} < \mathfrak{n}, \mathfrak{m} = \mathfrak{n}, \mathfrak{m} > \mathfrak{n},$$

where the last relation expresses $\mathfrak{n} < \mathfrak{m}$.

In order to say that any two cardinals satisfy *at least* one of these relations, two difficult questions must be answered:

(1) Do two sets M, N exist such that neither set can be put in a one-to-one correspondence with a subset of the other?

(2) Do two sets M, N exist such that each may be put in one-to-one correspondence with a proper subset of the other, but such that there is no one-to-one correspondence of all of the elements of M with all of the elements of N?

The second question is resolved by the so-called Cantor-Bernstein theorem, which proves that if M has the same cardinal number as a subset of N and N has the same cardinal number as a subset of M, then $\overline{\overline{M}} = \overline{\overline{N}}$.

The difficulty posed by the first question lies deeper, but it is overcome by showing that two such sets M and N do not exist. The following theorem results.

TRICHOTOMY THEOREM. *If $\mathfrak{m}$ and $\mathfrak{n}$ are any two cardinals, then exactly one of the relations $\mathfrak{m} = \mathfrak{n}$, $\mathfrak{m} < \mathfrak{n}$, $\mathfrak{m} > \mathfrak{n}$ is valid.*

Involved in the proof of this theorem is the celebrated assertion (introduced into mathematics in 1904 by the German mathematician, Ernst Zermelo, and called the *Axiom of Choice*):

If M is any collection of non-null sets which pairwise have no *elements in common, there exists at least one set N that has exactly one element in common with each set of the collection M.*

This statement was the cause of considerable controversy. Those who objected to it did so on the grounds that asserting the existence of a set for which no rule of construction is given is meaningless. It is not enough to say the set is "formed" by selecting an element from each set of the collection, for this might involve infinitely many *arbitrary* choices, and that is the very point at issue. Even today a number of very competent mathematicians refuse to make use of the axiom and do not accept as established a statement whose proof is accomplished only with its aid.

● EXERCISE

Assuming each infinite set contains a denumerable subset, show that $\aleph_0$ is the smallest transfinite cardinal.

II.6. Propositions

Logic is the subject that investigates, formulates, and establishes acceptable methods by which one statement may be asserted as a consequence of others. Everyone, non-scientist as well as scientist, reaches conclusions from premises (sometimes he even leaps at them), so the great importance of doing this in a manner that others recognize as valid is obvious.

Aristotle (384–322 B.C.) seems to have been the first to recognize logic as a separate discipline. In his *Organon* he refers to three "laws of thought" as being basic: (1) the *law of contradiction* (no proposition is both true and false), (2) *the law of identity* (each proposition implies itself), and (3) the *law of excluded middle* (each proposition is either true or false). Aristotle did not attempt to achieve for logic what his contemporary, Euclid, tried to accomplish for geometry—organize it into a deductive system.

We are concerned here with a small but important part of logic, known as the *propositional calculus*. The elements dealt with in that subject are propositions, and we shall define a proposition to be the meaning of a declarative sentence. Thus, "The woman is pretty" and "La mujer es bonita" are the same proposition, whereas "Oh, that I were wise" and "Give us this day our daily bread" are not propositions, since they do not assert something about something.

Propositions are frequently denoted by small letters, p, q, r, and so on, and are combined by the use of *logical connectives* to form other (compound) propositions. Thus, if p, q stand for propositions, they are combined by *disjunction* to form the proposition $p \lor q$ (read "p or q"), by *conjunction* to form the proposition $p \cdot q$ (read "p and q"), by *implication* to form the proposition $p \longrightarrow q$ (read "p implies q" or "if p then q"), by *equivalence* to form the proposition $p \equiv q$ (read "p is equivalent to q"), and from p is derived the proposition p', the *negation* of p (read "not p").

II.7. Truth Tables

Every proposition has a truth-value. In the classical (Aristotelian) logic, which nearly everyone still uses, every proposition is either true or false and not both (that is, it has exactly one of the truth-values T or F), whereas in some of the new logics propositions have more than two truth-values.

Expressions such as $p \lor q$, $p \cdot q$, $p \longrightarrow q$, in which the letters involved are regarded as *variables* whose "values" are propositions, are examples of *propositional functions*. They are not propositions as they stand, but they become propositions when specific propositions are substituted for the variables. Thus, if the proposition "Napoleon was Emperor of France" is substituted for p and "Less than two million people live in New York" is substituted for q, the propositional functions $p \lor q$, $p \cdot q$, $p \longrightarrow q$, p' become propositions. These propositional functions have, moreover, the property that the truth-value of each proposition obtained by substituting specific propositions for their variables depends only on the truth-values of those propositions. Such propositional functions are called *truth-functions*. An example of a propositional function that is *not* a truth function is "George believes that p," since if a specific proposition is substituted for p, the truth-value of the resulting proposition is not determined by the truth-value of the proposition substituted for p.

A truth-function that contains only a small number of variables may be conveniently analyzed by use of a truth table. The truth tables for the elementary truth-functions are as follows:

(I)		(II)			(III)			(IV)		
p	p'	p	q	$p \vee q$	p	q	$p \cdot q$	p	q	$p \longrightarrow q$
T	F	T	T	T	T	T	T	T	T	T
F	T	T	F	T	T	F	F	T	F	F
		F	T	T	F	T	F	F	T	T
		F	F	F	F	F	F	F	F	T

The entries in the last column of each table are the truth-values (T = true, F = false) of the truth-function (for which the table is constructed) corresponding to the indicated truth-values of the constituent propositions. Thus, by Table (I), the truth-function p' has truth-value T or F depending on whether p has truth-value F or T, respectively. Table (II) tells us that the disjunction $p \vee q$ is true *except when p and q are both false*. On the other hand, the conjunction $p \cdot q$ is false except when p and q are both true. All of this conforms to the way in which people ordinarily reason, though it might be noted that the first and fourth rows of Table (II) show that the "or" in "p or q" is used in the *inclusive* sense (that is, in the sense of and/or).

Table (IV), however, might seem rather surprising to the reader. From it we infer that if p and q are any given propositions, the proposition "p implies q" *is true in every case except when p is true and q is false*. The first and third rows of Table (IV) tell us that a true proposition is implied by any proposition (true or false), whereas the third and fourth rows show that a false proposition implies any proposition (true or false). To assert $p \longrightarrow q$ is, then, to assert that either p is false or q is true! This disjunction is frequently taken as the definition of implication.

The formal connection between the *antecedent* p and the *consequent* q, which is always present when the implication $p \longrightarrow q$ is asserted in ordinary discourse, is entirely lacking in the technical use of the term. Though the statements " 'New York is a tiny village' implies that 'Cleopatra was Queen of Egypt' and 'New York is a tiny village' implies that 'Cleopatra was not Queen of Egypt' " are both true [according to Table (IV)], they are not the sort of implications the layman would make.

All truth-functions can be expressed by using only the connectives of negation and disjunction. Calling two truth-functions the *same*, provided they yield the same proposition for the same set of values (that is, $T_1(p, q, \cdots, t) = T_2(p, q, \cdots, t)$, provided the proposition $T_1(p_0, q_0, \cdots, t_0)$ is the same (has the same meaning) as the proposition $T_2(p_0, q_0, \cdots, t_0)$, where $p_0, q_0, \cdots, t_0$ is any set of given propositions), we see, for example, that the truth-function $p \cdot q$ is the same as the truth-function $(p' \vee q')'$.

The reader can easily verify that all of the properties of set sum, product, and complementation given in Section II.3 are valid for proposition disjunction, conjunction, and negation, where the equality sign is interpreted as "is the same proposition as." By this is meant that if p, q, r denote given propositions, then

$$p \vee (q \vee r) = (p \vee q) \vee r,$$
$$p \vee q = q \vee p,$$
$$p \vee p = p,$$
$$p \cdot (q \cdot r) = (p \cdot q) \cdot r,$$
$$p \cdot q = q \cdot p,$$
$$p \cdot p = p,$$
$$p \cdot (q \vee r) = (p \cdot q) \vee (p \cdot r),$$
$$p \vee (q \cdot r) = (p \vee q) \cdot (p \vee r),$$
$$(p \vee q)' = p' \cdot q',$$
$$(p \cdot q)' = p' \vee q'.$$

The truth-function $p \equiv q$ is defined to be $(p \longrightarrow q) \cdot (q \longrightarrow p)$; that is, p is equivalent to q, provided each implies the other. Using Tables (III) and (IV), the truth-table for this connective is easily constructed,

p	q	$p \longrightarrow q$	$q \longrightarrow p$	$p \equiv q \equiv_D (p \longrightarrow q) \cdot (q \longrightarrow p)$
T	T	T	T	T
T	F	F	T	F
F	T	T	F	F
F	F	T	T	T

where "$\equiv_D$" means "is defined to be."

Note that $p \equiv q$ is true if and only if p and q are *both* true, or p and q are *both* false.

If the last column of the truth-table of a truth-function (containing only propositional variables and logical constants—a *formal* truth-function) consists only of T, the function is a *tautology;* if that column consists only of F, the function is *self-inconsistent,* and if the column contains at least one T and at least one F, the function is *contingent.* The five truth-functions whose truth tables are exhibited above are, therefore, all contingent. Examples of tautologies are: (1) $p \vee p'$, (2) $(p \cdot q) \longrightarrow p$, (3) $(p \longrightarrow q) \equiv (q' \longrightarrow p')$. The truth table for (3) is

p	q	p'	q'	$p \longrightarrow q$	$q' \longrightarrow p'$	$(p \longrightarrow q) \equiv (q' \longrightarrow p')$
T	T	F	F	T	T	T
T	F	F	T	F	F	T
F	T	T	F	T	T	T
F	F	T	T	T	T	T

The implication $q' \longrightarrow p'$ is called the *contrapositive* of the implication $p \longrightarrow q$. The logical equivalence of the two implications is very useful in mathematics, where a theorem is often established by proving its contrapositive.

● EXERCISES

1. Show that $p \cdot p'$ is self-inconsistent and that $[(p \longrightarrow q) \cdot p] \longrightarrow q$ and $[(p \longrightarrow q) \cdot q'] \longrightarrow p'$ are both tautologies.

2. Show that negation, disjunction, and conjunction may be defined in terms of one logical connective, denoted by / (called the Sheffer stroke function), interpreting p/q as "p is false or q is false."

II.8. Forms of Argumentation

Suppose it is admitted "If Smith is rich Smith is happy" and, further, it is agreed that Smith is rich. Then the conclusion, "Smith is happy," is regarded as valid. Stripped of its specific contents, the argument is formalized to yield the tautology $[(p \longrightarrow q) \cdot p] \longrightarrow q$. This argument, known in classical logic as *modus ponendo ponens*, is sometimes referred to as the *rule of affirming the antecedent*. The rule states that if an implication is asserted, and its antecedent is also asserted, the consequent may be detached from the implication and also asserted.

A truth-function T_1 is said to *necessarily imply* a truth-function T_2 if and only if the implication $T_1 \longrightarrow T_2$ is a tautology. Hence we may say that the basis of *modus ponendo ponens* is that the truth-function $(p \longrightarrow q) \cdot p$ necessarily implies the truth-function q.

Let us assume, again, the implication, "If Smith is rich, Smith is happy," is true, but suppose, now, Smith is not happy. Then the conclusion, "Smith is not rich," is regarded as valid. This is an example of another valid form of argumentation, *modus tollendo tollens*, or the *rule of denying the consequent*. Formalized, it rests on the fact that the truth-function $(p \longrightarrow q) \cdot q'$ necessarily implies the truth-function p'; that is, that the truth-function $[(p \longrightarrow q) \cdot q'] \longrightarrow p'$ is a tautology.

A common error is asserting the antecedent of an implication when the implication and its consequent are admitted. (If Jones is intelligent, Jones is honest. Jones is honest. Therefore Jones is intelligent.) Formalized, the truth-function $[(p \longrightarrow q) \cdot q] \longrightarrow p$ is *not* a tautology, so $[(p \longrightarrow q) \cdot q]$ does not necessarily imply p.

But though it is not valid in logic to affirm the antecedent when the implication and its consequent are true, if, for a large number of propositions $q_1, q_2, \cdots, q_n$ each of the conjunctions $(p \longrightarrow q_i) \cdot q_i$ $(i = 1, 2, \cdots, n)$ is true, the weight of evidence is in favor of p being true. For example, suppose it is granted "If Jones is the thief, Jones is six feet tall, has a wart on the end of his nose, speaks with a lisp, understands Swahili, and walks with a limp," and suppose each of the five consequents (to save space we have written one implication instead of five) is true. Then a jury might well be disposed to conclude that the antecedent, "Jones is the thief," is true. Such reasoning has much use in experimental science.

Another well-established procedure of experimental science is to formulate a "law" based on evidence obtained in the laboratory and then to seek verifications of the law by numerous experiments. After examining several samples of grass, for example, a scientist might assert "All grass is green" and then test this generalization by inspecting millions of specimens. A verification of the law is obtained whenever a sample of grass turns out to be green. But a blue moon is also a verification of this law! By the principle of contraposition, discussed in the preceding section, the assertion, "All grass is green," is logically equivalent to the assertion, "All not-green is not-grass," and a blue moon is a not-green that is a not-grass!

II.9. Deductive Theory

Let P denote a non-null abstract set whose elements are denoted by $p, q, r, \cdots$ and are called, for suggestiveness, *propositions*. Suppose P is *closed* with respect to a unary operation $'$, and a binary operation $\vee$; that is, corresponding to each element p of P, there is a unique element p' of P, and corresponding to each pair of elements p, q of P there is a unique element $p \vee q$ of P.

Any combination of elements of P and the operations $'$ and $\vee$ is a *truth-function* if and only if it conforms to the following requirements: (1) the "variables" $p, q, \cdots$ are truth-functions; (2) if A, B denote truth-functions, then so do A' and $A \vee B$; (3) no sequence of symbols is a truth-function unless it is formed according to (1) and (2) or unless it is replaceable, by definition, by a truth-function.

A certain subset of the class of truth-functions is called the class of tautologies. It is a remarkable fact that every tautology can be shown to follow from a set of *four* tautologies. *Defining* the binary operation $\longrightarrow$ ($p \longrightarrow q \equiv_D p' \vee q$), it is *assumed* that the truth-functions

$$(p \vee p) \longrightarrow p,$$
$$q \longrightarrow (p \vee q),$$
$$(p \vee q) \longrightarrow (q \vee p),$$
$$(q \longrightarrow r) \longrightarrow [(p \vee q) \longrightarrow (p \vee r)],$$

are tautologies. After rules are laid down that permit passing from one step in a proof to another, every tautology may be deduced from these four. It is interesting to observe that none of the three principles selected by Aristotle as fundamental laws of thought is included in these four basic tautologies. The Aristotelian principles are all deduced as theorems.

There are some misgivings concerning the suitability for mathematics of any logic such as that of Aristotle, in which the truth-function $p \vee p'$ is a tautology. Since, in the modern view, the abstractions dealt with in mathematics have no "real" existence, there would seem to be no basis for calling statements concerning them "true" or "false." The principal concern of mathematics is the validity (that is, *provability*) of its statements, but if the law of excluded middle $p \vee p'$ be interpreted to mean "For every proposition p, p is provable or its negation p' is provable," the law should probably be rejected in mathematics. Followers of the distinguished Dutch mathematician, L. E. J. Brouwer (1881–   ), do reject it. They use a logic, formulated by A. Heyting, in which $p \vee p'$ is *not* a tautology.

Postulational Systems

III.1. Undefined Terms and Unproved Propositions

"If you would converse with me," said Voltaire, "define your terms." This admonition is a reasonable one, provided it is not pushed too far. For if a conversation is to be fruitful, the participants must agree, as closely as possible, on the meaning of some of the basic terms they use, without attempting to define them. An attempt to define every term must lead to circularity (for example, a dictionary might define "exist" as "to be," and then define "be" as "to exist," with the result that "exist" means "to exist").

In a postulational system this difficulty is overcome by selecting certain concepts to be *primitive* or *undefined*, and then defining all additional notions (*peculiar to the subject*) in terms of those undefined ones. Thus the first step in forming a postulational system is *listing all the undefined terms*. As a practical convenience it is desirable to have only a few such terms, though reducing them to a minimum might result in undesirable complications. For example, we have seen that all the logical connectives can be defined by means of a single connective (the Sheffer stroke function), but elimination of all the other connectives would result in very cumbersome expressions for truth-functions.

If Voltaire had gone on to demand that anyone conversing with him prove everything they said, a similar difficulty would have been encountered. Participants in a discussion usually accept certain statements to be true without attempting to prove them; unless the speakers agree on some basic "principles" they can hardly agree on anything. But more important is the impossibility of proving *every* statement in the context of a given postulational system without arriving at a situation in which the proof of statement A contains statement B, whose proof, in turn, involves statement A.

So the second step in setting up a postulational system is *listing all the statements for which no proofs whatever are offered*. These statements are the postulates or axioms of the system. They state basic properties of the undefined, primitive notions, and they should be simple in structure and few in number. [The postulates can always be stated as one postulate by use of "and" (for example, P_1 and P_2 and $P_3 \cdots$ and P_n), but the postulate that results will not have a simple structure.]

The third step in forming a postulational system is extraneous, insofar as the initial basis is concerned, but is of intrinsic importance in developing the system. It consists of *laying down rules by means of which "new" statements can be deduced from "old" ones*. We shall call this step "choice of a logic." When no such choice is explicitly made it is assumed that the classical Aristotelian logic is to be used. All but a very few mathematicians never use any other logic, just as a century ago the only geometry considered was euclidean.

The fourth step in constructing a postulational system is *developing the system*. This consists of deducing the logical consequences of the postulates. They are the *theorems* of the system.

It is the unavoidable presence of undefined terms and unproved propositions in every postulational system that points up Russell's famous aphorism that in mathematics one never knows what one is talking about or whether what one is saying is true. Since the basic concepts are undefined, mathematicians do not know what they are talking about when they speak of these concepts (or of notions defined by means of them), and since all theorems depend

for their validity on unproved assumptions (the postulates), mathematicians do not know whether their assertions are really true.
Moreover, what does "true" mean when applied to a statement
containing undefined terms?

III.2. Consistency, Independence, and Completeness of a Postulational System

Postulational systems are rarely constructed just for the fun of
it. They are frequently formulated to exhibit a given branch of
mathematics (for example, geometry, set theory, group theory) as
a deductive system, and they are often suggested by a consideration of the common features of several different mathematical
structures. But even if an individual were moved to invent postulational systems for the mere pleasure of doing so, he would doubtless wish to develop the systems he creates, and if the development
were not to be entirely trivial, his systems would have to conform
to one requirement—*consistency.*

Since the basic concepts of a postulational system are undefined,
they are meaningless, and, consequently, the postulates of the system are propositional functions rather than propositions. When,
however, meanings are assigned to those concepts, the postulates
become propositions and hence have truth-values. A primitive
concept is said to be *interpreted* when a meaning is assigned to it,
and a *model* of the postulational system is obtained when each such
concept has been so interpreted that all the propositions arising
from the postulates are true.

We relate our notion of consistency of a postulational system Σ
to that of a model by defining Σ to be *consistent if and only if a
model of Σ exists.* This is the working definition that is usually
employed even when consistency is defined in another manner.
Assuming that mutually contradictory statements cannot hold in
a model, and that all interpretations of each consequence of Σ are
true statements in the respective models, it follows that if a model
of Σ exists, then it is not possible for both p and its negation p' to

be Σ-statements. (A Σ-statement is one that is either a postulate or a theorem of Σ.) This is frequently taken to be the definition of consistency. We assume the two notions to be equivalent.

Now the chief reason why *only* consistent postulational systems are of interest is that *any statement q is a Σ-statement in case Σ is not consistent.* For if Σ is not consistent, there is a Σ-statement p such that its negation p' is also a Σ-statement. Since p' is a Σ-statement, the implication $p \longrightarrow q$ is valid. But the validity of this implication, together with the fact that p is a Σ-statement, establishes (by *modus ponendo ponens*) the statement q. Hence inconsistent postulational systems are devoid of interest, and so it is of decisive importance to determine how a given postulational system behaves in that respect.

In order to assert that a given interpretation of the primitive concepts of a postulational system yields a model, one must have criteria for determining the truth of the particular propositions formed by the interpreted postulates. If, for example, one accepts as true the theorems of ordinary arithmetic, one postulational system (that of the real numbers) might serve as the model of another postulational system, which could then be said to be as consistent as the real number system. When Beltrami showed that the non-euclidean geometries can be interpreted as the geometry of certain surfaces in three-dimensional euclidean space he proved that those geometries are as consistent as euclidean geometry. Mathematicians are usually content with a postulational system that is as consistent as the real number system.

In the remainder of this section it will be assumed that Σ is a consistent postulational system.

A postulate P of Σ is called *independent* in Σ, provided the system $[\Sigma - (P)] + (P')$ is consistent (the notation indicates the system obtained from Σ on replacing postulate P by its negation P'). The system Σ is called independent if each of its postulates is independent in Σ.

If postulate P is an independent postulate in Σ, clearly it is not a logical consequence of the remaining postulates. If, on the other

hand, P is derivable from the other postulates of Σ, we may transfer
P from the list of postulates to the list of theorems. It is not a
serious defect in a postulational system to contain such *redundant*
postulates. In Hilbert's famous axiomatization of euclidean geometry there were postulates that were not independent. In Section
II.9 we listed four postulates on which the system of tautologies
is based. These postulates were given by the British philosopher-
logicians A. N. Whitehead and Bertrand Russell in their monu-
mental three-volume work, *Principia Mathematica*, the first volume
of which was published in 1910. These postulates form an independent set, but the system formulated in *Principia* contains a
fifth postulate,

$$[p \lor (q \lor r)] \longrightarrow [q \lor (p \lor r)],$$

which the German mathematician-logician Paul Bernays showed,
in 1926, to be deducible from the other four. Thus the system of
Whitehead and Russell was redundant.

The reader will recall from Chapter I that non-euclidean geometry owed its birth to the labors of those who hoped to show that
Euclid's system was redundant.

A consistent postulational system Σ is called logically *complete*,
provided the adjunction to Σ of any additional postulate (expressed
wholly in terms of either the primitive or the defined notions of Σ)
is either *unnecessary*, by virtue of the new postulate being a consequence of the postulates of Σ, or *unacceptable*, because the augmented system would be inconsistent.

It follows that if Σ is any logically complete postulational system,
and p denotes any statement involving only concepts of Σ, *either p
is derivable in Σ, or its negation p' is derivable in Σ*. For if p is *not*
a consequence of the postulates of Σ, then (since Σ is logically
complete) the augmented system $\Sigma + (p)$ is inconsistent; that is,
if p is assumed, a contradiction is encountered. In classical logic
this is sufficient to justify the assertion of p'.

Two models of a system Σ are isomorphic with respect to Σ, provided there exists a one-to-one correspondence between their elements and terms such that corresponding statements are interpretations of the same statement in Σ. If every two models of Σ are isomorphic with respect to Σ (and hence there is essentially only *one* model of Σ), the system is said to be *categorical*.

Every categorical system Σ is logically complete. For if Σ is not logically complete, there exists a statement p, involving only concepts of Σ, that is an *independent* postulate of the *consistent* system $\Sigma + (p)$. It follows that a model $\mathfrak{M}_1$ of system $\Sigma + (p)$ exists, as well as a model $\mathfrak{M}_2$ of system $\Sigma + (p')$. These two models are clearly *not* isomorphic, and each is a model of Σ, which contradicts the assumed categoricity of system Σ.

It is easily seen that, conversely, each non-categorical consistent system Σ is not logically complete, and so the concepts of logical completeness and categoricity are equivalent. The latter notion is occasionally thought of as a working definition of the former notion. We shall use that test in the next chapter to show that a given postulational system is logically complete.

It should be remarked that categoricity of a postulational system does not imply completeness of the system in the sense that every statement formulated in the concepts of the system may be *derived by means of a formalized theory of deductive procedure or that its negation may be so derived*. Thus, for example, postulates for the real number system are categorical, but there are propositions concerning real numbers which can be neither proved (within the system) nor disproved.

Though consistency of a postulational system is a *sine qua non*, and independence of the system is always desirable, the value of categoricity depends on the purpose the system is to serve. If the abstract form of one particular structure that has arisen in another way (for example, the real number system) is to be presented, the system must be categorical. But a non-categorical system represents features that numerous (non-isomorphic) mathematical structures have in common, and, consequently, its development results in the simultaneous investigation of those different structures. Such a system brings about a unification through generalization— one of the most important procedures of modern mathematics.

III.3. The Postulational System 7_3

The undefined terms of this system are a set P whose elements are called poins, and a set Λ whose elements are subsets of P and are called lins.

Postulate 1. *If $p, q \in P, p \neq q$, there is at least one $\lambda \in \Lambda$ such that $p \in \lambda$ and $q \in \lambda$.*
Postulate 2. *If $p, q \in P$, and $p \neq q$, there is at most one element λ of Λ such that $p \in \lambda$ and $q \in \lambda$.*
Postulate 3. *If $\lambda, \mu \in \Lambda, \lambda \neq \mu$, there is at least one element p of P such that $p \in \lambda$ and $p \in \mu$.*
Postulate 4. *The set Λ is not null.*
Postulate 5. *If $\lambda \in \Lambda$, there exist at least three elements p, q, r of P $(p \neq q \neq r \neq p)$ such that $p, q, r \in \lambda$.*
Postulate 6. *If $\lambda \in \Lambda$, there exists $p \in P$ such that $p \notin \lambda$.*
Postulate 7. *If $\lambda \in \Lambda$, there are at most three pairwise distinct elements of P that are elements of λ.*

Consistency of System 7_3

A model of the system is obtained by means of the following interpretations: the set P of poins is the set of seven letters A, B, C, D, E, F, G; and the set Λ of lins consists of those subsets of P that form the columns of the array

$$
\begin{array}{ccccccc}
A & B & C & D & E & F & G \\
B & C & D & E & F & G & A \\
D & E & F & G & A & B & C.
\end{array}
$$

Let the reader show that, with this interpretation, all seven postulates are true statements. Let $\mathfrak{M}_1$ denote this model.

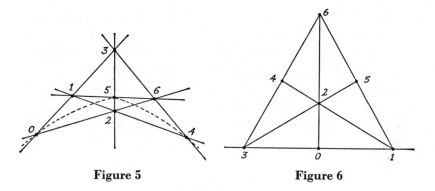

Figure 5   **Figure 6**

Figures 5 and 6 are geometric models of system 7₃. In each figure the poins are the points labeled 0, 1, 2, 3, 4, 5, 6, and the lins are the sets of collinear point-triples, *augmented by the triple* [0, 4, 5].

These three models of system 7₃ are pairwise isomorphic. This is easily seen by letting the letters A, B, C, D, E, F, G of the first model correspond, respectively, to the points 0, 1, 2, 3, 4, 5, 6 of either of the geometric models. Columnar triples of letters will then correspond to the lins of those models, with triple $[E, F, A]$ corresponding to triple [0, 4, 5] and all other columnar triples corresponding to collinear point-triples.

Independence of System 7₃

We shall exhibit seven models, in each of which the interpretation of six of the postulates and the negation of the remaining one are true statements.

INDEPENDENCE OF POSTULATE 1. Let the set P consist of the letters A, B, C, D, E, and let the set Λ consist of the triples $[A, B, C]$ and $[A, D, E]$. In this interpretation the *negation* of Postulate 1 is true, since there is no lin with poins B, D as elements, and Postulates 2, 3, 4, 5, 6, and 7 are all true.

INDEPENDENCE OF POSTULATE 2. Let the set P consist of the letters A, B, C, D, and let the set Λ consist of the triples $[A, B, C]$,

$[A, B, D]$, $[A, C, D]$. The negation of Postulate 2 is true, as are all the remaining postulates.

INDEPENDENCE OF POSTULATE 3.   Let the set P consist of the nine *marks* 1, 2, 3, 4, 5, 6, 7, 8, 9, and let the set Λ consist of the twelve triples $[1, 2, 3]$, $[4, 5, 6]$, $[7, 8, 9]$, $[1, 4, 7]$, $[2, 5, 8]$, $[3, 6, 9]$, $[1, 5, 9]$, $[3, 4, 8]$, $[2, 6, 7]$, $[3, 5, 7]$, $[1, 6, 8]$, $[2, 4, 9]$. (These twelve triples are seen to be the rows and columns of the array

$$
\begin{array}{ccc}
1 & 2 & 3 \\
4 & 5 & 6 \\
7 & 8 & 9
\end{array}
$$

augmented by the six triples obtained by developing the array as a determinant.) The negation of Postulate 3 and the Postulates 4, 5, 6, and 7 are obviously true in this interpretation. Let the reader show that Postulates 1 and 2 are also true.

INDEPENDENCE OF POSTULATE 4.   Let the set P consist of just one letter, A, and let the set Λ be the null set. The negation of Postulate 4 is clearly true in this interpretation. Now Postulate 1 asserts that if $p, q \in P$, $p \neq q$, there is at least one element of Λ that contains both of them. The reader will recall that an implication is true when the antecedent is false (see the truth table for *implication* in Section II.7). Hence the implication asserted by Postulate 1 is true, since its antecedent ($p, q \in P$, $p \neq q$) is false (because P consists of just one element). Similarly, it is shown that Postulates 2, 3, 5, 6, and 7 are true.

INDEPENDENCE OF POSTULATE 5.   Let the set P consist of the letters A, B, C, and let the set Λ consist of the pairs $[A, B]$, $[B, C]$, $[A, C]$.

INDEPENDENCE OF POSTULATE 6.   Let the set P consist of the letters A, B, C, and let the set Λ consist of the one triple $[A, B, C]$. Note that Postulate 3 is valid because its antecedent is false.

INDEPENDENCE OF POSTULATE 7.   Let the set P consist of all the *lines* of euclidean three-dimensional space E_3 that go through a point, and let the set Λ consist of all the planes of E_3 that pass through that same point. In this interpretation Postulates 1, 2, 3, 4, 5, and 6 are true, and Postulate 7 is false. Let the reader show this.

Observe that the model used to prove independence of Postulate 7 (the set of all lines and planes of E_3 that have a point in common) is itself part of a postulational system (the E_3), and so the system $7'_3$, formed by Postulates 1, 2, 3, 4, 5, 6 *and the negation of* Postulate 7, is only as consistent as the E_3. The following model is free from that criticism.

Let P consist of the thirteen consecutive letters of the English alphabet from A to M, and let Λ consist of those quadruples of letters that form the columns of the array

$$
\begin{array}{ccccccccccccc}
A & B & C & D & E & F & G & H & I & J & K & L & M \\
B & C & D & E & F & G & H & I & J & K & L & M & A \\
D & E & F & G & H & I & J & K & L & M & A & B & C \\
J & K & L & M & A & B & C & D & E & F & G & H & I.
\end{array}
$$

Clearly, Postulate 7 is false in this interpretation, and Postulates 4, 5, and 6 are true. It is easy to verify that Postulates 1, 2, and 3 are also true.

Since the two models for the system $7'_3$ are *not* isomorphic (in one of them P has infinitely many elements, whereas in the other P has exactly thirteen elements), that system is *not* categorical. We shall show, on the other hand, that system 7_3 is categorical.

LEMMA III.3.1.   *If p, $q \in P$, $p \neq q$, there is exactly one element λ of Λ such that p, $q \in \lambda$.*

This is an immediate consequence of Postulates 1 and 2.

LEMMA III.3.2.   *If λ, $\mu \in \Lambda$, $\lambda \neq \mu$, there is exactly one element p of P such that $p \in \lambda$ and $p \in \mu$.*

This follows from Postulate 3 and Lemma III.3.1.

LEMMA III.3.3. *There are three pairwise distinct elements of P that are not elements of the same* λ, λ ∈ Λ.

Proof. According to Postulates 4 and 5, P has pairwise distinct elements, denoted by A, B, D, that belong to one element of Λ, which, according to Postulate 7, can be denoted by $[A, B, D]$. Use of Postulate 6 implies the existence of an element C of P which is not an element of $[A, B, D]$. Then elements A, B, C are pairwise distinct and are not elements of the same element of Λ (Lemma III.3.1).

LEMMA III.3.4. *The set P contains at least seven elements.*

Proof. Consider the pairwise distinct elements denoted by A, B, C, D, shown to exist by Lemma III.3.3, with $[A, B, D] \in \Lambda$ and $C \notin [A, B, D]$. According to Lemma III.3.1, B, C determine a unique element of Λ which contains exactly one element of P, say E, distinct from B and C. Let $[B, C, E]$ denote that element of Λ. Using Lemma III.3.1 again gives $E \neq A$ and $E \neq D$, so elements A, B, C, D, E are pairwise distinct. In a similar way, elements C, D imply the existence of an element F of P with $[C, D, F] \in \Lambda$ and the six elements A, B, C, D, E, F pairwise distinct, whereas D, E imply the existence of an element G of P with $[D, E, G] \in \Lambda$ and the seven elements A, B, C, D, E, F, G pairwise distinct.

LEMMA III.3.5. *The set P contains exactly seven elements.*

Proof. Lemma III.3.4 established the existence of seven pairwise distinct elements, denoted by A, B, C, D, E, F, G, and four elements of Λ, $[A, B, D]$, $[B, C, E]$, $[C, D, F]$, and $[D, E, G]$.

Suppose P contains an element H, distinct from each of these seven elements. Then A, H determine an element of Λ whose third constituent we denote by X. Applying Lemma III.3.2 to $[A, H, X]$ and to $[B, C, E]$, $[C, D, F]$, and $[D, E, G]$, in turn, gives $X = C$ or $X = E$ *and* $X = C$, or $X = F$ *and* $X = E$ or $X = G$. These conditions are obviously impossible to satisfy, so no eighth element H exists.

LEMMA III.3.6. *The system 7_3 is isomorphic to the model $\mathfrak{M}_1$.*

Proof. According to Lemma III.3.5, P consists of seven elements, denoted by A, B, C, D, E, F, G, and $[A, B, D]$, $[B, C, E]$, $[C, D, F]$, $[D, E, G]$ denote elements of Λ. It is easily seen that the third constituents of those elements of Λ determined by (E, F), (F, G), (G, A) are A, B, C, respectively. Hence Λ consists of the four elements given above and the three additional ones, $[E, F, A]$, $[F, G, B]$, $[G, A, C]$. If, therefore, *the elements of P be made to correspond to their labels, the desired isomorphism results.*

THEOREM III.3.1. *The postulational system 7_3 is categorical.*

Proof. The preceding argument shows that any model of system 7_3 is isomorphic to $\mathfrak{M}_1$.

COROLLARY III.3.1. *The postulational system 7_3 is complete.*

The postulational system 7_3 defines a (finite) *projective geometry of seven points and seven lines* (to drop the terms poin and lin, which were intended to emphasize the abstract nature of the elements); each point is on three lines, and each line is on three points (see Figure 5).

III.4. A Finite Affine Geometry

The finite projective geometry developed in the preceding section is a member of a class of plane geometries whose coordinatizations are treated at length in later chapters. In this section we give the axiomatization of a particular finite affine geometry—an example of the kind of plane geometry discussed in Chapters IV and V.

The undefined terms are an abstract set S of elements called points, and a set L of subsets of S called lines.

Postulate 1. *If a, b are points, $a \neq b$, there is at least one line containing both of them.*

Postulate 2. *If a, b are points, $a \neq b$, there is at most one line containing both of them.*

Postulate 3. *If p denotes any point, and l denotes any line, with $p \notin l$, there is at least one line that contains p and has no point in common with l.*

Postulate 4. *If p denotes any point, and l denotes any line, with $p \notin l$, there is at most one line that contains p and has no point in common with l.*

Postulate 5. *Every line contains at least three pairwise distinct points.*

Postulate 6. *Not all elements of S belong to the same line.*

Postulate 7. *The set L has at least one element.*

Postulate 8. *No line contains more than three pairwise distinct points.*

The model used to prove the independence of Postulate 3 of the system 7_3 (Section III.3) is also a model of this system. (Let the reader verify this.) Arguments similar to those used to establish Theorem III.3.1 show that any two models of this system are isomorphic, so the system characterizes the affine geometry of nine points and twelve lines, each line containing exactly three points and each point being on exactly four lines. The system is independent and categorical (and hence complete).

● EXERCISES

In the geometry defined by the postulational system of this section, show that

1. Every line has exactly two lines parallel to it.
2. The six points of two parallel lines, joined consecutively in proper order, determine a hexagon such that the intersections of opposite sides are collinear (Pappus-Pascal theorem).

III.5. Hilbert's Postulates for Three-dimensional Euclidean Geometry

We conclude this chapter with a very brief account of the postulates for three-dimensional euclidean geometry formulated by the German mathematician, David Hilbert. In their original formulation (1898–1899) these postulates did not constitute an independent set. This defect was removed by the American, E. H. Moore (1902), the German, Arthur Rosenthal (1912), and others, and simplifications of the system were made by Rosenthal and Bernays.

Numerous postulational systems for euclidean geometry are now available, many of which have certain advantages over Hilbert's system. In a later chapter we shall discuss one such system that arose as the result of a very new approach to geometry. But Hilbert's system is still the best known to mathematicians. Its importance is not confined to the successful accomplishment of its objective, but is due also to the fact that the great prestige of its creator stimulated many other gifted individuals to cultivate that part of mathematics. The period from 1880 to 1910 saw the publication of 1,385 articles devoted to the foundations of geometry, and there is still much activity in that field.

An English translation of Hilbert's *Grundlagen der Geometrie* is available (Open Court, 1902), and the simplified system of postulates is readily accessible in that language (see, for example, Harold Wolfe, *Non-euclidean Geometry*, Dryden, 1945, pp. 12–16).

It suffices for our purpose to note that Hilbert postulated three sets of undefined elements called, respectively, *points, lines,* and *planes*. There are, in addition, three primitive relations called *incidence, betweenness* or *order,* and *congruence*.

Hilbert's postulates are divided into five groups. *Group I (Postulates of Connection)* consists of eight postulates concerning the incidence relation. Here we encounter such assertions as, "If A, B are any two distinct points, there exists at least one line that is incident with them both," and "Each line is incident with at least two distinct points."

Group II (Axioms of Order) consists of four postulates that express properties of the undefined notion of betweenness. The reader will recall that one of the defects of Euclid's *Elements* is that betweenness notions for points on a line were used without any postulational justification. We owe the introduction of postulates of betweenness into geometry, and a systematic study of them, to the German mathematician, Moritz Pasch (1882). Hilbert assumes, for example, that (1) if point B is between points A and C, the three points A, B, C are pairwise distinct, are incident with the same line, and B is between C and A, and (2) if A, C are distinct points, there exists an element B of the line determined by A, C such that C is between A and B. Two more postulates complete this group.

Group III (*Postulates of Congruence*) consists of five axioms. They accomplish what Euclid accomplished through his tacit introduction of the vague principle of superposition. It is assumed, for example, that two triangles are congruent if two sides and the included angle of one are congruent, respectively, to two sides and the included angle of the other.

Group IV (*Parallel Postulate*) consists of just one postulate, the parallel postulate in the form given by Playfair (see Section I.5).

Group V (*Postulates of Continuity*) consists of two postulates. The first (the so-called *Axiom of Archimedes*) asserts that if AB and CD are any two segments (a notion defined in terms of betweenness), there is a positive integer n and n points $A_1, A_2, \cdots, A_n$ on the line incident with the points A, B, such that each of the segments $AA_1, A_1A_2, \cdots, A_{n-1}A_n$ is congruent to segment CD and B is between A and A_n.

The second postulate of Group V (the *Completeness Axiom*) was the subject of much controversy. Bernays replaced the original form of this postulate by an *a priori* less-sweeping assertion (the linear completeness axiom), which states that no additional points can be added to the points incident with a line so that all the other postulates are valid in the extended system.

The two postulates of continuity endow the line and the circle, for example, with the kind of structure that enables us to give a rigorous proof for the construction of an equilateral triangle, which Euclid failed to do in Proposition 1, Book I of his *Elements* (see Section I.2).

Hilbert's postulational system is as consistent as the arithmetic of real numbers.

Of special interest is the independence of Postulate IV (the parallel postulate). If that postulate be deleted from the set, the postulational system that remains defines what Bolyai called *absolute geometry*. It is the common core of the euclidean and hyperbolic (non-euclidean) geometries, for the latter is obtained by adjoining to the postulates of absolute geometry the hyperbolic form of the parallel postulate (for example, if l is any line and p is any point not on l, the plane determined by p and l contains at least two lines through p, neither of which intersects l), and the former results when the euclidean parallel postulate is put back in the system.

Absolute geometry is a good example of a postulational system that is very rich in consequences and which is easily proved to be incomplete. Consider the proposition, "The angle-sum of every triangle equals two right angles." This is *not* provable in absolute geometry, for if it were, it would be provable in hyperbolic geometry (since all the postulates of absolute geometry are also postulates of hyperbolic geometry), and of course in hyperbolic geometry the angle-sum of every triangle is *less than* two right angles. On the other hand, the negation of the proposition (that is, there exists at least one triangle whose angle-sum is different from two right angles) is also not provable in absolute geometry. If it were, it would be provable in euclidean geometry (since all the postulates of absolute geometry are also postulates of euclidean geometry), and in euclidean geometry the angle-sum of *every* triangle equals two right angles. Hence the proposition is *undecidable* in absolute geometry, which is, therefore, incomplete.

IV

Coordinates in an
Affine Plane

Foreword

We have observed that the epoch-making discoveries of the young Bolyai were published in an appendix to the scholarly but prosaic treatise on geometry written by his father. Another appendix to a book that was of incomparably greater significance than the book itself was the first treatise on analytic geometry, which formed an appendix to *Discours de la Methode*, written by the French philosopher-mathematician, René Descartes (1596–1650), and published at Leyden in 1637. It would be difficult to overestimate the value, for the progress of mathematics, of the union of geometry and algebra that was initiated by Descartes' contribution. The basis for that union is the establishment of a coordinate system. This is easily accomplished in elementary analytic geometry by assuming that the points of a line are in a one-to-one correspondence with the numbers of the real number system and that the space being coordinatized (line, plane, three-space) has all of the euclidean properties.

In this chapter the procedure shall be very different from that. We start with a rudimentary (affine) plane (which might contain only a finite number of points), and assign, as coordinates to points, ordered pairs of elements of an *abstract* "coordinate set." Assuming additional postulates for the plane results in additional properties of the coordinate set, and we finally obtain a set that is isomorphic to the real numbers.

IV.1. The Affine Plane II

The primitive notions of II are an abstract set Σ of elements (for suggestiveness, called points), and certain subsets of Σ (called lines), which are subjected to the following postulates.

Postulate 1. *If $P, Q \in \Sigma$, $(P \neq Q)$, there is one and only one line of which both P and Q are elements.*

If a point P is an element of a line, we shall say that the line contains the point, the point is on the line, or the line is on the point.

THEOREM IV.1.1. *If p, q denote any two distinct lines, there is at most one point of Σ that is on both p and q.*

Two distinct lines with a point in common are called *intersecting* and are said to *intersect;* if they do not intersect, they are mutually parallel. If two lines are mutually parallel, each is said to be parallel to the other, and vice versa.

Postulate 2. *If $P \in \Sigma$, and p denotes any line such that $P \notin p$, there is exactly one line on P that is parallel to p.*

Postulate 3. *There exists at least one quadruple of pairwise distinct points, no three of which are on the same line.*

The postulational system defined by Postulates 1, 2, and 3 is consistent, for a model of the system is easily obtained by interpreting the pointset Σ as the set $[A, B, C, D]$ of four letters, and the lines as the pairs (A, B), (A, C), (A, D), (B, C), (B, D), (C, D). Note that each "line" has the same number of "points" on it and each "point" is on the same number of "lines" (the number being one greater than the number of points on any one line). That this homogeneity is not peculiar to the model, but is present in every plane II, is shown by the following theorem.

THEOREM IV.1.2. *Every line of* II *contains the same number of points.*

Proof. The postulates insure that there is a line p, a point P, $P \notin p$, and exactly one line on P that is parallel to p. Since every other line on P intersects p in exactly one point, and each point of p determines with P a unique line on P, there is a one-to-one correspondence between the points on p and the lines on P that intersect p. Hence the number of points on p is one less than the number of lines on P.

Now consider a set of four pairwise distinct points, P, Q, R, S, no three of which are collinear (that is, are contained in the same line). The existence of at least one such quadruple is insured by Postulate 3. Let n denote the number of pairwise distinct points on the unique line containing Q and R. (We speak of this line as being *determined* by Q, R, and denote it by line QR.) Note that n may denote either a finite or a transfinite cardinal number.

Since $P \notin$ line QR and $P \notin$ line RS, each of these lines contains exactly one less point than the number of lines on P, and, consequently, line RS contains exactly n points also. (If n is transfinite, the reader may easily establish a one-to-one correspondence between the points of line QR and the points of line RS.) Consideration of point Q and lines PS and RS shows that line PS contains exactly n points; in a similar way it is seen that each of the six lines determined by points P, Q, R, S (taken in pairs) contains exactly n points.

Now let q denote any line of II. If $P \in q$, either q is one of the six lines determined by the four points P, Q, R, S—and con-consequently contains exactly n points—or it is not one of those lines. In the latter event, $Q \notin q$, so q contains the same number of points as does line RS (according to the first part of the proof). Finally, if $P \notin q$, then q contains the same number of points as does line QR (that is, n points), and the proof is complete.

COROLLARY IV.1.1. *Every point of* Π *has the same number of lines on it. This number is one more than the number of points on any one line of* Π.

Remark. By Postulate 3 and the above corollary, there are at least three lines on every point of Π and at least two points on every line of Π.

● EXERCISES

1. Show that if a line of Π contains exactly n points, Π contains exactly n^2 points and $n(n + 1)$ lines.
2. Show that if a line is parallel to one of two intersecting lines, it intersects the other.

IV.2. Parallel Classes

Note that Postulate 2 is the euclidean parallel postulate in Playfair's form. Because of this, some writers refer to plane Π as euclidean instead of affine.

It is clear that the relation of parallelism is *symmetric* (that is, if line p is parallel to line q, line q is parallel to line p), but *not reflexive* (that is, line p is not parallel to itself), since parallelism is a relation of one line to another (distinct) line.

If p, q, r are pairwise distinct lines such that p is parallel to q, and q is parallel to r, then p is parallel to r. For if the contrary be assumed, p and r have exactly one point P in common, since they are distinct lines. According to the symmetry of parallelism, there are two distinct lines on P, each parallel to line q, in contradiction to Postulate 2. Hence for *three pairwise distinct* lines, parallelism is transitive. The restriction of pairwise distinctness cannot be omitted, since parallelism is *not* reflexive.

DEFINITION. If p is any line of Π, those lines of Π consisting of p and all lines parallel to p form the *parallel class of p*. This class is denoted by $[p]$.

The following remarks are easily established:

(i) If a line of Π contains exactly n points, class $[p]$ has exactly n members, for each line p of Π.

(ii) If $q \in [p]$, then $[q] = [p]$.

(iii) Two parallel classes are either mutually exclusive (that is, have no common member) or coincide.

IV.3. Coordinatizing the Plane Π

To establish a coordinate system in Π the following arbitrary choices are made.

(1) Select *any* point of Π, label it O, and refer to it as the *origin*.

(2) Select *any* three lines on O, refer to any one of them as the x-line, another one as the y-line, and the third one as the *unit line*.

(3) Select *any* point on the unit line, different from O, label it I, and refer to it as the *unit point*.

(4) Select *any* abstract set whose cardinality is the same as the cardinality of the set of points on the unit line. Denote this abstract set by Γ, refer to it as the *coordinate set*, and let γ denote an arbitrarily chosen (but fixed) one-to-one correspondence between the points of the unit line and the elements of Γ. Label those elements of Γ that correspond by γ to the points O, I of the unit line by 0, 1, respectively. We write $0 = \gamma(O)$, $1 = \gamma(I)$, and emphasize that the symbols 0, 1 are merely labels for two *distinct* elements of the abstract set Γ. In this context they do *not* stand

for those real numbers they are usually employed to represent. Nothing is assumed about the character of the elements of the coordinate set Γ. The reader may think of them in any way he chooses—as watermelons or lamb chops—just so each is distinguishable from each of the others.

We now associate with each point of Π an ordered pair of elements of Γ as coordinates, in the following manner.

(1) Let A denote any point of the unit line OI, and let $a = \gamma(A)$; that is, a denotes that unique element of Γ that corresponds to point A by means of the one-to-one correspondence γ. Then the pair (a, a) is assigned to point A as its coordinates. It follows that O has coordinates $(0, 0)$ and I has coordinates $(1, 1)$.

Note that if $x \in \Gamma$, there is a unique point X of line OI such that $x = \gamma(X)$, and, consequently, the coordinates of X are (x, x).

(2) Let P denote any point of Π that is *not* on the unit line OI. The unique line on P in the parallel class of the y-line intersects the unit line (see Exercise 2, Section IV.1) in point A, and the unique line on P in the

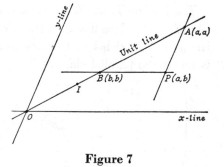

Figure 7

parallel class of the x-line intersects the unit line in point B. If A has coordinates (a, a) and B has coordinates (b, b) (in step 1 coordinates were assigned to each point of the unit line), the pair (a, b) is assigned to P as its coordinates.

As in elementary analytic geometry, the first coordinate of a point is called its *abscissa*, and the second coordinate its *ordinate*.

Note. In the many figures we shall use, lines and points will be represented in the conventional way. But the reader should never lose sight of the *abstract* nature of the subject and should not infer more from the diagrams than they are intended to convey. For example, the solid lines of the figures do not imply that the lines

contain infinitely many points. Also, there is no significance to the
angle made by the x- and y-lines in the drawings. The reader may
draw these lines always mutually perpendicular if he wishes to
do so.

● EXERCISES

1. Show that each point on the x-line has coordinates of the form $(x, 0)$
and each point of the y-line has coordinates of the form $(0, y)$.

2. Show that if a, b are any elements of Γ, there is exactly one point
of Π with coordinates (a, b).

IV.4. Slope and Equation of a Line

The line on I, in the parallel class of the y-line, is called the
line of slopes. This line is distinct from the y-line, since it contains
the point I, which is not on the y-line. Now if p is any line that
is *not* in the parallel class of the y-line, let p^* denote the line on

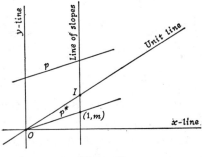

point O, in the parallel class
of line p. Line p^* intersects
the line of slopes in a point
whose coordinates are $(1, m)$,
and the element m of Γ is assigned to line p as *slope*.
Lines in the parallel class
of the y-line are not assigned
slopes.

Figure 8

Remark. If lines p, q are
mutually parallel, and line p
has slope m, then line q has slope m. This follows directly from
the method of assigning slopes to lines if either p or q is on point
O, and if neither line is on O, the transitivity of parallelism (for
pairwise distinct lines) gives the desired result.

Conversely, if two distinct lines have the same slope, they are
mutually parallel.

The x-line has slope 0, and the unit line has slope 1.

An equation of a line is any equation that is satisfied by the coordinates of every point on the line, and which is *not* satisfied by the coordinates of any point that is not on the line.

Clearly, an equation of the x-line is $y = 0$, since a point is on the x-line if and only if its ordinate is 0 (let the reader show this), and each line in the parallel class of the x-line has equation $y = b$, for some element b of Γ. Similarly, each line in the parallel class of the y-line has equation $x = a$, for some element a of Γ. The y-line itself has equation $x = 0$, and $x = 1$ is an equation of the line of slopes. The unit line has equation $y = x$, since a point is on the unit line if and only if its ordinate and its abscissa are the same element of Γ (let the reader show this).

If a line intersects the y-line in a unique point $(0, b)$, the ordinate b is called the y-intercept of the line. We have seen that equations of certain lines of II are easily written, and they have the same forms as in elementary analytic geometry, but we are not yet in a position to write an equation of a line with y-intercept b and slope m, for arbitrary elements m, b of Γ. This is accomplished in the following section by defining a ternary operation in the coordinate set Γ.

● EXERCISE

Construct the unique line with y-intercept b and slope m.

IV.5. The Ternary Operation *T*

We shall now establish a procedure for associating with every ordered triple a, m, b of elements of Γ a unique element of Γ, denoted by $T(a, m, b)$.

Consider the line p with slope m and y-intercept b. The line on point $(a, 0)$, in the parallel class of the y-line, intersects p (let the reader show this), and the ordinate of that point of intersection is defined to be $T(a, m, b)$.

If $P(x, y)$ is an arbitrary point on line p, clearly $y = T(x, m, b)$. On the other hand, if $P(x, y)$ is *not* on line p, let Q denote the

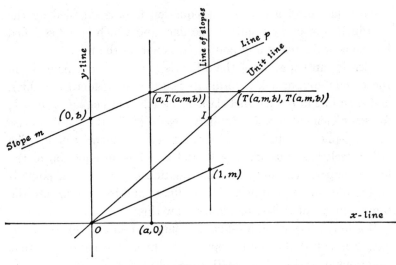

Figure 9

intersection of p with the line on P, in the parallel class of the y-line. Since $P \neq Q$, and each of these points has the same abscissa x, then (in view of the one-to-one correspondence between the points of Π and the ordered pairs of elements of Γ) the ordinate y of P is different from the ordinate $T(x, m, b)$ of Q; that is, $y \neq T(x, m, b)$. We conclude, therefore, that $y = T(x, m, b)$ *is an equation of the line with slope m and y-intercept b.*

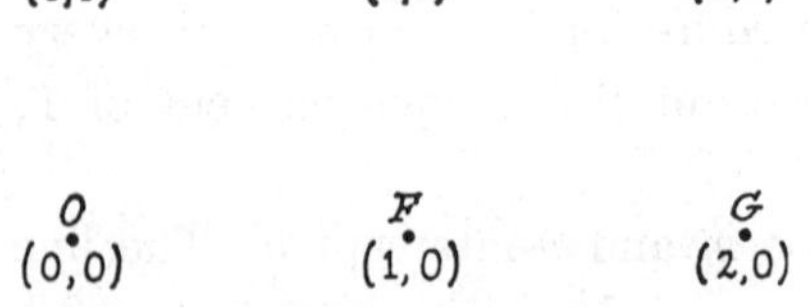

Example. Consider the plane Π of nine points—denoted by the letters A, B, C, D, E, F, G, I, O—whose lines are the vertical and horizontal triples of Figure 10, together with the six triples AIG, CDF, BEO, CIO, AEF, and BDG.

Let Γ consist of the three marks 0, 1, 2. Select ODA as the

A
$(0,2)$

B
$(1,2)$

C
$(2,2)$

D
$(0,1)$

I
$(1,1)$

E
$(2,1)$

O
$(0,0)$

F
$(1,0)$

G
$(2,0)$

Figure 10

y-line, OFG as the x-line, OIC as the unit line, and I as the unit point. Then O, I, and C have coordinates $(0, 0)$, $(1, 1)$, and $(2, 2)$, respectively. Since lines ODA, FIB, GEC form the parallel class of the y-line, and lines OFG, DIE, ABC form the parallel class of the x-line, our method of associating coordinates with points gives the points the coordinates shown in Figure 10.

Line FIB is the line of slopes. Slope is undefined for each of the lines ODA, FIB, GEC (constituting the parallel class of the y-line), and it is 0 for each of the lines OFG, DIE, ABC. Line OBE contains O and the point B (of the line of slopes) with coordinates $(1, 2)$. Hence line OBE has slope 2. Since each of the lines AIG, CDF is parallel to line OBE, each also has slope 2. Finally, line CIO intersects the line of slopes in point $I(1, 1)$, and, consequently it, together with lines AEF, DBG, has slope 1.

This information is tabulated on the right. Let us now compute a few values of the ternary operator. To evaluate $T(0, 2, 1)$ we seek the line with y-intercept 1 and slope 2, that is, the line on D with slope 2. The table gives line CDF. The line on $(0, 0)$ in the parallel class of the y-line (this is the y-line itself) meets line CDF in point D. Since the ordinate of D is 1, $T(0, 2, 1) = 1$.

To find $T(1, 2, 2)$ we intersect the line on $(1, 0)$, in the parallel class of the y-line (that is, the line FIB), with the line that has slope 2 and y-intercept 2 (that is, the line AIG). The intersection I has ordinate 1, and, consequently, $T(1, 2, 2) = 1$.

Line	Slope
ODA	Undefined
FIB	Undefined
GEC	Undefined
ABC	0
DIE	0
OFG	0
AEF	1
DBG	1
CIO	1
OBE	2
AIG	2
CDF	2

● EXERCISES

1. Evaluate $T(2, 2, 2)$, $T(1, 1, 1)$, $T(2, 1, 1)$.
2. Write equations of lines AEF, AIG.

IV.6. The Planar Ternary Ring [Γ, T]

The ternary operator $T(a, m, b)$, defined geometrically in the preceding section, gives the abstract coordinate set Γ an algebraic structure by virtue of which it is called a *planar ternary ring.* (Sometimes it is called a Hall planar ternary ring, after its inventor, the American mathematician, Marshall Hall—the "planar" is due to its genesis.) We shall list and prove the pertinent properties, the first two of which are immediate.

(1) Γ contains distinct elements 0 *and* 1.

(2) If a, b, c ∈ Γ, *then* $T(a, b, c)$ *is a unique element of* Γ.

(3) For all a, b, c ∈ Γ, $T(0, b, c) = T(a, 0, c) = c.$ The line on $(0, 0)$, in the parallel class of the y-line (that is, the y-line itself), intersects the line with slope b and y-intercept c in a point whose ordinate is c [so $T(0, b, c) = c$], whereas the line on $(a, 0)$, in the parallel class of the y-line, intersects the line with slope 0 and y-intercept c in a point whose ordinate is c [so $T(a, 0, c) = c$].

(4) For every element a of Γ, $T(a, 1, 0) = T(1, a, 0) = a.$ $T(a, 1, 0)$ is the ordinate of the intersection of the unit line with the line on $(a, 0)$, in the parallel class of the y-line, and so equals a, whereas $T(1, a, 0)$ is the ordinate of the intersection of the line of slopes with the line on O that has slope a; consequently, $T(1, a, 0) = a.$

(5) If m, m′, b, b′ ∈ Γ, $m \neq m′$, *the equation* $T(x, m, b) = T(x, m′, b′)$ *has a unique solution in* Γ. If p is the line with slope m and y-intercept b, and q is the line with slope $m′$ and y-intercept $b′$, the condition $m \neq m′$ implies that p, q are distinct and not mutually parallel. Hence they have a unique point $P(a, c)$ in common, and $c = T(a, m, b) = T(a, m′, b′)$; that is, the element a of Γ is the unique solution of the equation $T(x, m, b) = T(x, m′, b′)$.

(6) If a, a′, b, b′ ∈ Γ, $a \neq a′$, *the system of equations*

$$T(a, x, y) = b,$$
$$T(a′, x, y) = b′$$

has a unique solution in Γ. Since $a \neq a′$, the unique line p on the two points (a, b), $(a′, b′)$ is not in the parallel class of the

y-line. Consequently, p has a slope m and a y-intercept k; that is, $T(a, m, k) = b$ and $T(a', m, k) = b'$.

(7) For all $a, m, c \in \Gamma$, the equation $T(a, m, x) = c$ has a unique solution in Γ. Let p denote the unique line on the point (a, c) in the parallel class of the line joining O to the point $(1, m)$, and denote the y-intercept of p by b. Then $c = T(a, m, b)$.

IV.7. The Affine Plane Defined by a Ternary Ring

In the preceding section we saw how an affine plane Π gives rise to a planar ternary ring whose elements form the coordinate set Γ of Π (that is, there is a one-to-one correspondence between the points of Π and ordered pairs of elements of Γ). The question arises: If $[\Gamma, T]$ is *any* ternary ring (that is, *any abstract set Γ and ternary operation T, defined in Γ, having properties 1–7*, as given in *Section IV.6*), is there an affine plane Π with coordinate set Γ and planar ternary ring $[\Gamma, T]$? *The answer is yes,* and points and lines are defined as follows.

DEFINITIONS. Let Σ denote the set of all ordered pairs of elements of Γ. These are the *points* of Π; that is, Σ is the pointset of Π.

A *line of the first kind* of Π, denoted by $[a]$, $a \in \Gamma$, is the set of all points (a, y), $y \in \Gamma$.

A *line of the second kind* of Π, denoted by $[m, b]$, $m, b \in \Gamma$, is the set of all points $(x, T(x, m, b))$, $x \in \Gamma$.

Two lines are parallel if and only if they have no point in common.

In order to show that the points and lines just defined form a plane Π, Postulates 1, 2, and 3, Section IV.1, so interpreted, must be proved true. We establish first some lemmas.

LEMMA IV.7.1. *For all $a, m, b \in \Gamma$, line $[a]$ intersects line $[m, b]$; that is, each line of the first kind intersects each line of the second kind.*

Proof. The line $[a]$ is the set of all points (a, y), $y \in \Gamma$, and line $[m, b]$ is the set of all points $(x, T(x, m, b))$, $x \in \Gamma$. It is clear that the point $(a, T(a, m, b))$ belongs to both sets.

LEMMA IV.7.2. *Two distinct lines $[m, b]$, $[m', b']$ intersect if and only if $m \neq m'$.*

Proof. If $m \neq m'$, the equation $T(x, m, b) = T(x, m', b')$ has a unique solution (property 5, Section IV.6), say $x = a$. Then the point $(a, T(a, m, b))$ is on both of the lines $[m, b]$, $[m', b']$.

If $[m, b]$ and $[m', b']$ intersect, a point (c, d) exists such that $(c, d) \in [m, b]$ and $(c, d) \in [m', b']$; that is, $d = T(c, m, b)$ and $d = T(c, m', b')$. Now by property 7 (Section IV.6), the equation $d = T(c, m, x)$ has a unique solution. If $m = m'$, both b and b' are solutions of that equation, and, consequently, $m = m'$ implies $b = b'$. But this is impossible, since the lines $[m, b]$, $[m', b']$ are assumed to be distinct. Hence $m \neq m'$, and the lemma is proved.

LEMMA IV.7.3. *Two lines are mutually parallel if and only if they are of the forms $[a]$, $[b]$, $a \neq b$, or of the forms $[m, b]$, $[m, b']$, $b \neq b'$.*

Proof. If the lines are of the forms $[a]$, $[b]$, $a \neq b$, they are obviously parallel, and if they are of the forms $[m, b]$, $[m, b']$, $b \neq b'$, they are parallel by the preceding lemma.

On the other hand, if p, p' denote two mutually parallel lines, by Lemma IV.7.1 they are of the forms $[a]$, $[b]$, $a \neq b$, or $[m, b]$, $[m', b']$. In the latter case, it follows from Lemma IV.7.2 that $m = m'$, and, consequently, $b \neq b'$.

THEOREM IV.7.1. *If (a, c) is a point not on a line p, there is one and only one line on (a, c) parallel to p.*

Proof. If line p is of the form $[d]$, then $a \neq d$, since $(a, c) \notin [d]$. Then line $[a]$ contains (a, c) and is parallel to $[d]$ (Lemma IV.7.3). Now $[a]$ is the only line of the first kind that contains point (a, c), and by Lemma IV.7.1 no line of the second kind is parallel to $[d]$.

Hence, in this case, there is a unique line on (a, c) parallel to line p.

If line p is of the form $[m, b]$, each line parallel to $[m, b]$ is of the form $[m, b']$, $b \neq b'$, by Lemma IV.7.3. Now by property 7 (Section IV.6) the equation $c = T(a, m, x)$ has a unique solution, say $x = b^*$. Then $c = T(a, m, b^*)$, and, consequently, $b^* \neq b$, for, since $(a, c) \notin [m, b]$, $c \neq T(a, m, b)$. Hence the line $[m, b^*]$, $b^* \neq b$, contains point (a, c), is parallel to line $[m, b]$, and is the only line with these properties.

THEOREM IV.7.2. *If (a, b), (a', b') are distinct points, there is exactly one line containing both points.*

Proof. If $a = a'$, then $b \neq b'$. Clearly line $[a]$ contains both points, and no other line of the first kind does. No line of the form $[m, c]$ contains both points, since if $b = T(a, m, c)$, then $b' \neq T(a', m, c) = T(a, m, c)$.

If $a \neq a'$, the system of equations

$$T(a, x, y) = b,$$

$$T(a', x, y) = b'$$

has a unique solution (property 6, Section IV.6), say $x = m$, $y = b^*$. Then the line $[m, b^*]$ contains (a, b) and (a', b') and is the only line of the second kind to do so. Since $a \neq a'$, no line of the form $[d]$ contains both points, and the proof is complete.

THEOREM IV.7.3. *There exists at least one quadruple of pairwise distinct points, no three of which are on the same line.*

Proof. The four points $(0, 0)$, $(0, 1)$, $(1, 0)$, $(1, 1)$ have the desired property. By the proof of the preceding theorem, points $(0, 0)$ and $(0, 1)$ are contained in line $[0]$ and in no other line, and points $(1, 0)$, $(1, 1)$ are contained in line $[1]$ and in no other line. Neither of the points $(0, 0)$, $(0, 1)$ is on line $[1]$, and neither of the points $(1, 0)$, $(1, 1)$ is on line $[0]$. Hence no three of these four points are collinear.

These three theorems establish that an affine plane Π arises from any given ternary ring $[\Gamma, T]$ on defining the points of the plane to be ordered pairs of elements of the abstract set Γ, and the lines to be: (1) subsets of points of the form (a, y), a being an arbitrary but fixed element of Γ and $y \in \Gamma$ (denoted by $[a]$), and (2) subsets of points of the form $(x, T(x, m, b))$, m, b being arbitrarily chosen elements of Γ and $x \in \Gamma$ (denoted by $[m, b]$). Clearly, Γ is the coordinate set of Π, $(0, 0)$ being the origin O, line $[0, 0]$ being the x-line, line $[0]$ being the y-line, line $[1, 0]$ being the unit line, point $(1, 1)$ being the unit point I, and line $[1]$ being the slope line.

Let $T^*(a, m, b)$ be the ternary operation defined in Π (with respect to the coordinate system just described) in the manner of Section IV.5. Then $T^*(a, m, b)$ is the ordinate of the intersection of the lines $[m, b]$ and $[a]$. Since $[m, b]$ is the class of all points $(x, T(x, m, b))$, $x \in \Gamma$, and $[a]$ is the class of all points (a, y), $y \in \Gamma$, the intersection of these lines is the point $(a, T(a, m, b))$. Consequently, $T^*(a, m, b) = T(a, m, b)$, so the ternary operation in the ring $[\Gamma, T]$ is identical to the ternary operation in the associated affine plane Π. Thus the abstract equivalence between the "geometric" entity *affine plane* and the "algebraic" entity *ternary ring* is fully established.

IV.8. Introduction of Addition

We now define a binary operation (denoted by $+$) for points of the unit line, which associates with each ordered pair of those points a third point, called their (ordered) *sum*, and we refer to the operation as *addition*. This induces an addition in the ternary ring $[\Gamma, T]$.

Let $A(a, a)$, $B(b, b)$ be points of the unit line (Figure 11). Intersect the line on B, in the parallel class of the x-line, with the y-line, obtaining the point $(0, b)$, and intersect the line on $(0, b)$, in the parallel class of the unit line, with the line on A, in the parallel class of the y-line, obtaining the point $P(a, c)$. The line on P, in the parallel class of the x-line, intersects the unit line in the point $C(c, c)$. Define $C = A + B$, and $c = a + b$.

Remark 1. For every point A of the unit line, $A + O = A = O + A$, and, consequently, for every element a of Γ, $a + 0 = a = 0 + a$. Let the reader establish this.

Remark 2. The expression $a + b = c (a, b, c \in \Gamma)$ uniquely determines any one of the elements a, b, c when the other two are given.

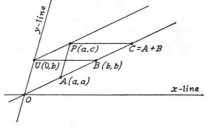

Figure 11

Given a and b, the definition of addition determines c uniquely. Suppose a and c are given, and let P denote the intersection of the line on $A(a, a)$, in the parallel class of the y-line, with the line on $C(c, c)$, in the parallel class of the x-line. The line on P, in the parallel class of the unit line, intersects the y-line in the point $(0, b)$, with $a + b = c$. (Let the reader draw a figure and verify this.) Finally, if b and c are given, a is similarly uniquely determined.

Let $[\Gamma, +]$ denote the coordinate set Γ, (ordered) addition being defined for each two elements of Γ as above. Since $0 \in [\Gamma, +]$ such that $a + 0 = a = 0 + a$, for every element a, and the expression $a + b = c$ uniquely determines any one of the three elements when the other two are given, $[\Gamma, +]$ forms (with respect to $+$) an algebraic structure known as a *loop*.

Remark 3. If we examine the construction that defines the sum $a + b$, we see that $a + b$ is the *ordinate* of the point of intersection of the line with slope 1 and y-intercept b, with the line on $(a, 0)$, in the parallel class of the y-line. Turning to the definition of the ternary operator T, we see that the ordinate of that intersection is $T(a, 1, b)$, and, consequently, for all elements a, b of Γ,

$(*)$ $$a + b = T(a, 1, b).$$

Remark 4. Since $y = T(x, 1, b)$ is an equation for the line with slope 1 and y-intercept b, it follows from (∗) that such a line has equation $y = x + b$, exactly as in elementary analytic geometry.

IV.9. Introduction of Multiplication

We define a binary operation called *multiplication* (denoted by ·) for all ordered pairs of points of the unit line, which induces a multiplication in the ternary ring $[\Gamma, T]$.

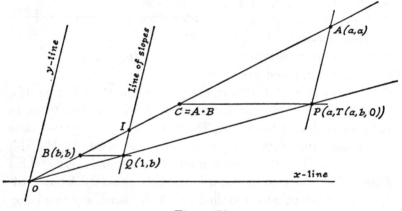

Figure 12

Let $A(a, a)$, $B(b, b)$ be points of the unit line OI (Figure 12). The line on B, in the parallel class of the x-line, intersects the line of slopes in the point Q having coordinates $(1, b)$. The line OQ intersects the line on A, in the parallel class of the y-line, in the point P, and the line on P, in the parallel class of the x-line, intersects the unit line in point C. Define $C = A \cdot B$, and, if C has coordinates (c, c), put $c = a \cdot b$.

Remark 1. For every point A of the unit line, $A \cdot 0 = 0 = 0 \cdot A$, and hence for every element a of Γ, $a \cdot 0 = 0 = 0 \cdot a$. Let the reader establish this.

Remark 2. If $A, B, B' \in$ line OI, $A \neq 0$, $A \cdot B = A \cdot B'$, then

$B = B'$. Let A, B, B' have coordinates (a, a), (b, b), (b', b'), respectively, and suppose $B \neq B'$ (Figure 13).

Then $b \neq b'$, $Q(1, b) \neq Q'(1, b')$, and line $OQ \neq$ line OQ'. Hence the respective intersections P, P' of these lines with the line on A, in the parallel class of the y-line, are distinct (since $A \neq 0$). The coordinates of P and P' are $(a, a \cdot b)$ and $(a, a \cdot b')$, respectively, and $P \neq P'$ implies $a \cdot b \neq a \cdot b'$, and, consequently, $A \cdot B \neq A \cdot B'$. Hence, if $A \neq 0$, $B \neq B'$ implies $A \cdot B \neq A \cdot B'$, which is logically equivalent to Remark 2.

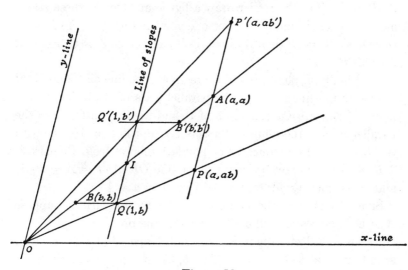

Figure 13

Remark 3. If A, A', $B \in$ line OI, $B \neq 0$, $A \cdot B = A' \cdot B$, then $A = A'$. The proof is similar to that of Remark 2.

Remark 4. If A, $B \in$ line OI, $A \neq 0$, $B \neq 0$, then $A \cdot B \neq 0$. By Remark 2, $A \neq 0$ and $B \neq 0$ imply $A \cdot B \neq A \cdot 0 = 0$, by Remark 1.

COROLLARY. If A, $B \in$ line OI, then $A \cdot B = 0$ if and only if either $A = 0$ or $B = 0$. Also, if a, $b \in \Gamma$, then $a \cdot b = 0$ if and only if $a = 0$ or $b = 0$.

Remark 5. If A, B, $C \in$ line OI, $A \neq O$, $B \neq O$, $C \neq O$, then $A \cdot B = C$ uniquely determines any one of the points A, B, C when the other two are given.

(1) Point C is uniquely determined by the definition of $A \cdot B$.

(2) If $A(a,a)$ and $C(c,c)$ are given, each different from O the line on C, in the parallel class of the x-line, intersects the line on A, in the parallel class of the y-line, in the point $Q(a,c)$, $a \neq 0$. The line OQ meets the line of slopes (since $Q \notin y$-line) in a point $(1, b)$, for some element b of Γ. Then the point $B(b, b)$ is such that $A \cdot B = C$. (Let the reader draw a figure and verify these statements.) There is, moreover, just one point B such that $A \cdot B = C$, since if $A \cdot B' = C$, then $A \cdot B = A \cdot B'$, and since $A \neq O$, Remark 2 yields $B = B'$.

(3) Finally, if points B and C are given, the line on O and $(1, b)$ intersects the line on C, in the parallel class of the x-line, in the point (a, c), for some element a of Γ, and the line on (a, c), in the parallel class of the y-line, intersects the unit line in $A(a, a)$, with $A \cdot B = C$. The uniqueness of point A follows from Remark 3.

Remark 6. For every point A of line OI, $A \cdot I = A = I \cdot A$, and, consequently, for every $a \in \Gamma$, $a \cdot 1 = a = 1 \cdot a$.

Since for $A = O$ the remark is valid by Remark 1, suppose $A \neq O$. The reader will easily complete the proof.

Remark 7. Referring to Figure 12, if $A(a, a)$, $B(b, b)$, $C(c, c)$ are such that $C = A \cdot B$, then $c = T(a, b, 0)$, and, consequently, multiplication in Γ is expressible in terms of the ternary operation T; that is, $a \cdot b = T(a, b, 0)$ for all elements a, b of Γ. Now any line on O, with slope m, has equation $y = T(x, m, 0)$, so each such line has equation $y = x \cdot m$, exactly as in elementary analytic geometry.

Remark 8. By virtue of Remarks 5 and 6, the set Γ' (read "Γ apostrophe") forms a *loop with respect to multiplication*, where Γ' denotes the set obtained by deleting from Γ the element 0. We denote this loop by $[\Gamma', \cdot]$.

IV.10. Vectors

DEFINITIONS. An ordered pair A, B of points of Π is called a *vector* and is denoted by $\overrightarrow{AB}$. Point A is the first or initial point of vector $\overrightarrow{AB}$, and point B is its second or terminal point. Any line containing the initial and the terminal points of a vector is called a *carrier* of the vector. If $A = B$, vector $\overrightarrow{AB}$ is called a *null* vector.

The first problem raised by the introduction of vectors is the important one of defining an *equivalence relation* for them. This is any binary *reflexive, symmetric,* and *transitive* relation (see Section IV.2) with respect to which each two vectors are comparable. Using the ordinary sign of equality to denote such a relation for vectors, we lay down the following criteria.

RULE 1. *A null vector is equivalent to a vector if and only if that vector is also null; that is, $\overrightarrow{AA} = \overrightarrow{CD}$ if and only if $C = D$.*

RULE 2. *Each vector is equivalent to itself; that is, $\overrightarrow{AB} = \overrightarrow{AB}$, for every vector $\overrightarrow{AB}$.*

RULE 3. *If the carriers of two non-null vectors $\overrightarrow{AB}$, $\overrightarrow{CD}$ are distinct lines, $\overrightarrow{AB} = \overrightarrow{CD}$ if and only if line AB is parallel to line CD and line AC is parallel to line BD.*

Remark. If $\overrightarrow{AB}$, $\overrightarrow{CD}$ have distinct carriers, $\overrightarrow{AB} = \overrightarrow{CD}$ implies $\overrightarrow{AC} = \overrightarrow{BD}$.

These rules do not suffice to establish the desired equivalence relation. First, they give no criterion for equivalence of two non-null vectors with the *same* carrier, but a less obvious difficulty arises even for vectors with distinct carriers. Do these three rules insure the *transitivity* of the relation they seek to define, for three non-null vectors with pairwise distinct carriers? If $\overrightarrow{AB}$, $\overrightarrow{A'B'}$,

$\overrightarrow{A''B''}$ are three non-null vectors, with pairwise distinct carriers, and $\overrightarrow{AB} = \overrightarrow{A'B'}$, $\overrightarrow{A'B'} = \overrightarrow{A''B''}$, by virtue of the criteria given in Rules 1, 2, and 3, does it necessarily follow that $\overrightarrow{AB} = \overrightarrow{A''B''}$? In view of Rule 3, this question is equivalent to the following one: If lines AB, $A'B'$, $A''B''$ are pairwise mutually parallel, and line AA' is parallel to line BB', and line $A'A''$ is parallel to line $B'B''$, does it follow that line AA'' is parallel to line BB''?

This question is affirmatively answered for the ordinary euclidean plane studied in high school geometry (and in the usual college analytic geometry course), but in distinction to the many postulates required for the ordinary euclidean plane (in Hilbert's system, Section III.5, fifteen of the twenty postulates refer to the plane alone), our rudimentary plane Π is defined by only three postulates. Now it is a fact (which will be established in the next section) that planes Π (satisfying Postulates 1, 2, and 3, Section IV.1) exist for which the last question posed above must be answered in the negative! Such planes Π will be eliminated from consideration by means of an additional postulate.

IV.11. A Remarkable Affine Plane Π

The object of this section is to give an example of an affine plane Π [that is, a set of points and lines (subsets of the set of points) satisfying Postulates 1, 2, and 3, Section IV.1] that contains two triples A, A', A'' and B, B', B'' of pairwise distinct points such that the lines AB, $A'B'$, $A''B''$ are pairwise mutually parallel, line AA' is parallel to line BB', and line $A'A''$ is parallel to line $B'B''$, but line AA'' is *not* parallel to line BB''.

The points of Π are the points of the ordinary euclidean plane of high school or of elementary analytic geometry. The lines of Π are: (1) the vertical euclidean straight lines, $x = $ constant; (2) the horizontal euclidean straight lines, $y = $ constant; (3) the euclidean straight lines with *negative* slopes (the "halves" of such lines that lie on or above the x-axis make any angle between 90° and 180°

with the positive direction of the x-axis); and (4) the euclidean
broken lines defined by the equations

$$y = (\tfrac{1}{2})(x - a) \tan \theta, \quad y \geqq 0,$$
$$y = (x - a) \tan \theta, \quad y < 0,$$

for every real number a and every angle θ, $0° < \theta < 90°$ (see Figure 14).

We show now that the
points and lines defined
above have the properties
stated by Postulates 1, 2,
and 3 of Section IV.1, and,
consequently, form an affine
plane II.

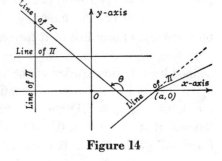

Figure 14

*Each two distinct points
$A(a_1, a_2)$, $B(b_1, b_2)$ of II are on
exactly one line.*

Proof. This is clear if A, B belong to the same half-plane
(bounded by the x-axis), or if the euclidean straight line joining
them does not have a positive slope. If neither alternative holds,
the labeling is assumed such that $A(a_1, a_2)$ is in the upper half-
plane (that is, $a_2 > 0$) and $B(b_1, b_2)$ is in the lower half-plane
(that is, $b_2 < 0$). Since the line AB has a positive slope, $a_1 > b_1$.
There is a unique line on points A, B if and only if the equations

$$a_2 = (\tfrac{1}{2})(a_1 - a) \tan \theta,$$
$$b_2 = (b_1 - a) \tan \theta,$$

have a unique solution (a, θ), with $0° < \theta < 90°$. Solving, we
obtain $a = (2a_2 b_1 - a_1 b_2)/(2a_2 - b_2)$ (since $a_2 > 0$ and $b_2 < 0$,
the denominator $2a_2 - b_2 > 0$) and $\tan \theta = (2a_2 - b_2)/(a_1 - b_1)$.
Since $a_1 > b_1$, the denominator of the last fraction is positive, so
$\tan \theta > 0$. Hence the equations possess a unique solution (a, θ),
with $0° < \theta < 90°$, and, consequently, A, B are on a unique line
of II in all cases.

If p is any line and P any point of Π, *P not on p, there is a unique line of* Π *on P that is parallel to p.*

Proof. Suppose, first, p is a euclidean line. Clearly, every *vertical* line and every *horizontal* line of Π intersects every line of Π formed by a broken euclidean line. Let us prove that also every euclidean line with negative slope has this property.

Let $y = mx + b$, $m < 0$, be such a line, and let

(∗) $\qquad y = (\tfrac{1}{2})(x - a) \tan \theta, \quad y \geqq 0,$

$\qquad\qquad y = (x - a) \tan \theta, \quad y < 0, \quad 0° < \theta < 90°$

be equations of any line of Π formed by a broken euclidean line.

If $y = mx + b$ fails to meet the upper half of the line of Π defined by (∗), substitution of the value of x given by $mx + b = (\tfrac{1}{2})(x - a) \tan \theta$ in $y = mx + b$ must result in a negative value for y; that is, $y = [(\tfrac{1}{2})(am + b) \tan \theta]/[(\tfrac{1}{2}) \tan \theta - m] < 0$. It follows that $am + b < 0$. But then $y = mx + b$ meets the *lower* half of the line of Π defined by (∗) in the point whose coordinates are

$\qquad x = (a \tan \theta + b)/(\tan \theta - m),$

$\qquad y = (am + b) \tan \theta / (\tan \theta - m) < 0.$

Hence, if p is a vertical (horizontal) line, the unique vertical (horizontal) line on P that is parallel to p (in the euclidean sense) is the desired line.

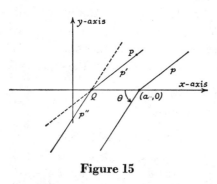

Figure 15

Say p is a line of Π given by $y = (\tfrac{1}{2})(x - a) \tan \theta$, $y \geqq 0$; $y = (x - a) \tan \theta$, $y < 0$, and $0° < \theta < 90°$. If P is in the upper half-plane, let p' denote the unique euclidean straight line on P, parallel to the upper half of line p, and denote by Q the intersection of p' with the x-axis (Figure 15). If p'' denotes the euclidean line on Q, parallel to the lower half of line p, the line of Π formed by the upper half of p' and the lower half of p'' is surely parallel to p.

No other line on P has this property. From what has been established in the first part of this proof, it suffices to show that each broken line $y = (\frac{1}{2})(x - b) \tan \varphi, y \geqq 0; y = (x - b) \tan \varphi,$ $y < 0, 0° < \varphi < 90°, \varphi \neq \theta$, intersects line p. The first two equations in each set are satisfied by the pair of numbers

$$x_0 = (a \tan \theta - b \tan \varphi)/(\tan \theta - \tan \varphi),$$
$$y_0 = [(\tfrac{1}{2})(a - b) \tan \theta \tan \varphi]/(\tan \theta - \tan \varphi),$$

and the second two equations are satisfied by the pair $x_0, 2y_0$. Hence the two upper halves of the lines intersect when $y_0 \geqq 0$, and the two lower halves meet when $y_0 < 0$. In either event, the lines are not mutually parallel.

If $P(x,y)$ has $y \leqq 0$, a similar procedure is employed.

There exists at least one quadruple of pairwise distinct points, no three of which are on the same line.

Proof. The points $(0, 0)$, $(1, 0)$, $(0, 1)$, $(1, 1)$ are easily seen to form such a quadruple.

Now consider the three pairwise mutually parallel lines AB, $A'B'$, $A''B''$ of plane II, shown in Figure 16, line AA' being par-

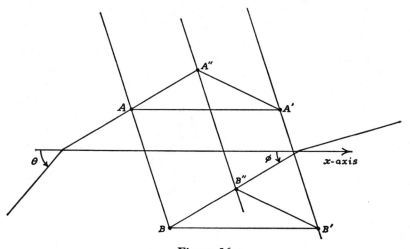

Figure 16

allel to line BB' (they are each horizontal euclidean lines) and line $A'A''$ being parallel to line $B'B''$ (they are mutually parallel euclidean lines). The lines AA'' and BB'' are *not* parallel; since $\theta > \varphi$, the two lower halves of these two lines intersect, as a direct consequence of Euclid's fifth postulate!

Let the reader establish algebraically that $\theta > \varphi$.

V

Coordinates in an Affine Plane with Desargues and Pappus Properties

Foreword (The First Desargues Property)

For a reason that will be made clear later, an affine plane is said to have *the first Desargues property*, provided (whenever lines AB, $A'B'$, $A''B''$ of the plane are pairwise mutually parallel) that when line AA' is parallel to line BB' and line $A'A''$ is parallel to line $B'B''$, then line AA'' is parallel to line BB''. Girard Desargues (1593–1662), the French architect-mathematician, was one of the first to study conic sections as projections of circles.

The property enunciated above is a special case of his beautiful theorem concerning perspective triangles to which we shall often refer. The example of an affine plane given in Section IV.11 shows that not every affine plane has the first Desargues property. Since this property is essential in order that the binary relation for vectors we are establishing be an equivalence relation, we restrict our attention from now on to those planes that, in addition to satisfying Postulates 1, 2, and 3, Section IV.1, have the first Desargues property. In these planes the equivalence relation for three non-null vectors with pairwise distinct carriers is transitive. The ordinary euclidean plane has the first Desargues property.

V.1. Completion of the Equivalence Definition for Vectors

The definition of the desired relation is now completed by defining it for distinct vectors with the same carrier. The criterion is embodied in the following rule.

RULE 4.   *If $\overrightarrow{AB}$, $\overrightarrow{EF}$ are distinct vectors with the same carrier, $\overrightarrow{AB} = \overrightarrow{EF}$ if and only if a vector $\overrightarrow{CD}$ exists, with carrier distinct from that of $\overrightarrow{AB}$, such that (under Rule 3) $\overrightarrow{AB} = \overrightarrow{CD}$ and $\overrightarrow{CD} = \overrightarrow{EF}$.*

Clearly, we wish the equivalence of the vectors $\overrightarrow{AB}$, $\overrightarrow{EF}$, with the same carrier, to be *independent* of the choice of the auxiliary vector $\overrightarrow{CD}$, so Rule 4 needs to be justified by showing that this is indeed the case.

Let $\overrightarrow{C'D'}$ be another choice of a vector, with carrier distinct from that of $\overrightarrow{AB}$, such that $\overrightarrow{AB} = \overrightarrow{C'D'}$ and $\overrightarrow{C'D'} = \overrightarrow{EG}$. Suppose $\overrightarrow{CD}$ and $\overrightarrow{C'D'}$ have different carriers (see Figure 17).

Since the carriers of vectors $\overrightarrow{AB}$, $\overrightarrow{CD}$, $\overrightarrow{C'D'}$ are pairwise distinct, and the plane has the first Desargues property, $\overrightarrow{AB} = \overrightarrow{C'D'}$, $\overrightarrow{AB} = \overrightarrow{CD}$ imply $\overrightarrow{CD} = \overrightarrow{C'D'}$. This, together with $\overrightarrow{CD} = \overrightarrow{EF}$,

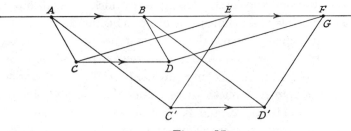

Figure 17

yields (for the same reason) $\overrightarrow{C'D'} = \overrightarrow{EF}$, and, consequently, line $D'F$ is parallel to line $C'E$. But $\overrightarrow{C'D'} = \overrightarrow{EG}$ implies line $D'G$ is parallel to line $C'E$. Hence line $D'F$ is identical to line $D'G$, so $F = G$.

Thus, if $\overrightarrow{AB} = \overrightarrow{EF}$ follows from Rule 4, by use of a vector $\overrightarrow{CD}$ whose carrier is distinct from that of $\overrightarrow{AB}$, $\overrightarrow{AB} = \overrightarrow{EF}$ follows also, by use of *any* vector $\overrightarrow{C'D'}$ whose carrier is distinct from that of $\overrightarrow{CD}$ and such that $\overrightarrow{AB} = \overrightarrow{C'D'}$, $\overrightarrow{EF} = \overrightarrow{C'D'}$. How would the reader proceed in the event a vector $\overrightarrow{C'D'}$ has the same carrier as $\overrightarrow{CD}$?

● EXERCISES

1. Show that the binary relation defined by Rules 1–4 in the set of vectors of plane II, with the first Desargues property is an equivalence relation (that is, is reflexive, symmetric, and transitive).

2. If $\overrightarrow{AA'} = \overrightarrow{BB'}$, $\overrightarrow{A'A''} = \overrightarrow{B'B''}$ (no two vectors having the same carriers), prove $\overrightarrow{AA''} = \overrightarrow{BB''}$. How is this property related to the first Desargues property?

3. Prove $\overrightarrow{AB} = \overrightarrow{AC}$ if and only if $B = C$.

V.2. Addition of Vectors

If $\overrightarrow{AB}$ is any vector and P is any point, there is exactly one point Q such that $\overrightarrow{AB} = \overrightarrow{PQ}$. There is at most one such point Q, since if $\overrightarrow{AB} = \overrightarrow{PQ_1}$ and $\overrightarrow{AB} = \overrightarrow{PQ_2}$, then $\overrightarrow{PQ_1} = \overrightarrow{PQ_2}$, by transitivity of vector equivalence, and $Q_1 = Q_2$ (Exercise 3, Section V.1). If $\overrightarrow{AB}$ is a null vector, $Q = P$ is the desired point (Rule 1, Section IV.10). If $\overrightarrow{AB}$ is not a null vector, suppose, first, P is not on the carrier of $\overrightarrow{AB}$. The line on P, parallel to line AB, intersects the line on B, parallel to line AP, in the desired point Q ($\overrightarrow{AB} = \overrightarrow{PQ}$, by Rule 3, Section IV.10). Finally, if P is on the line AB, let $\overrightarrow{CD}$

be any vector such that $\overrightarrow{AB} = \overrightarrow{CD}$ and line AB is distinct from line CD. The line on D, parallel to line CP, meets line AB in the desired point Q ($\overrightarrow{AB} = \overrightarrow{PQ}$, by Rule 4, Section V.1).

DEFINITION (VECTOR ADDITION). *Let $\overrightarrow{AB}, \overrightarrow{CD}$ be any two vectors. If $B = C$, then $\overrightarrow{AB} + \overrightarrow{CD} = \overrightarrow{AD}$. If $B \neq C$, and Q denotes the unique point such that $\overrightarrow{BQ} = \overrightarrow{CD}$, then $\overrightarrow{AB} + \overrightarrow{CD} = \overrightarrow{AQ}$.*

Remark 1. If $\overrightarrow{CD} = \overrightarrow{C'D'}$, $\overrightarrow{AB} + \overrightarrow{CD} = \overrightarrow{AB} + \overrightarrow{C'D'}$. For, by definition of vector addition, $\overrightarrow{AB} + \overrightarrow{CD} = \overrightarrow{AQ} = \overrightarrow{AB} + \overrightarrow{BQ}$, where $\overrightarrow{BQ} = \overrightarrow{CD}$. Then $\overrightarrow{BQ} = \overrightarrow{C'D'}$, and $\overrightarrow{AB} + \overrightarrow{C'D'} = \overrightarrow{AB} + \overrightarrow{BQ} = \overrightarrow{AQ}$.

Remark 2. If $\overrightarrow{CD} = \overrightarrow{C'D'}$, $\overrightarrow{CD} + \overrightarrow{AB} = \overrightarrow{C'D'} + \overrightarrow{AB}$. For $\overrightarrow{CD} + \overrightarrow{AB} = \overrightarrow{CD} + \overrightarrow{DQ} = \overrightarrow{CQ}$, where $\overrightarrow{DQ} = \overrightarrow{AB}$ and $\overrightarrow{C'D'} + \overrightarrow{AB} = \overrightarrow{C'D'} + \overrightarrow{D'Q'} = \overrightarrow{C'Q'}$, where $\overrightarrow{D'Q'} = \overrightarrow{AB}$. Hence $\overrightarrow{DQ} = \overrightarrow{D'Q'}$, which, together with $\overrightarrow{CD} = \overrightarrow{C'D'}$, gives $\overrightarrow{CQ} = \overrightarrow{C'Q'}$. This is justified by Exercise 2, Section V.1, if the carriers of $\overrightarrow{CD}, \overrightarrow{C'D'}, \overrightarrow{DQ}, \overrightarrow{D'Q'}$ are pairwise distinct. Let the reader check the other possibility.

Remark 3. If $\overrightarrow{AB} = \overrightarrow{A'B'}$ and $\overrightarrow{CD} = \overrightarrow{C'D'}$, $\overrightarrow{AB} + \overrightarrow{CD} = \overrightarrow{A'B'} + \overrightarrow{C'D'}$. For, by Remark 2, $\overrightarrow{AB} + \overrightarrow{CD} = \overrightarrow{A'B'} + \overrightarrow{CD}$, and, by Remark 1, $\overrightarrow{A'B'} + \overrightarrow{CD} = \overrightarrow{A'B'} + \overrightarrow{C'D'}$. Transitivity of vector equivalence gives $\overrightarrow{AB} + \overrightarrow{CD} = \overrightarrow{A'B'} + \overrightarrow{C'D'}$.

THEOREM V.2.1. *Vector addition is associative.*

Proof. If $\overrightarrow{AB}, \overrightarrow{CD}, \overrightarrow{EF}$ are any three vectors, consider the unique vectors $\overrightarrow{BP}, \overrightarrow{PQ}$ such that $\overrightarrow{BP} = \overrightarrow{CD}$ and $\overrightarrow{PQ} = \overrightarrow{EF}$. Then

$$(\overrightarrow{AB} + \overrightarrow{CD}) + \overrightarrow{EF} = (\overrightarrow{AB} + \overrightarrow{BP}) + \overrightarrow{PQ} \quad \text{(Remark 3)}$$
$$= \overrightarrow{AP} + \overrightarrow{PQ} \quad \text{(definition of vector addition)}$$
$$= \overrightarrow{AQ} \quad \text{(definition of vector addition)},$$

$$\overrightarrow{AB} + (\overrightarrow{CD} + \overrightarrow{EF}) = \overrightarrow{AB} + (\overrightarrow{BP} + \overrightarrow{PQ})$$
$$= \overrightarrow{AB} + \overrightarrow{BQ}$$
$$= \overrightarrow{AQ},$$

and, consequently, $(\overrightarrow{AB} + \overrightarrow{CD}) + \overrightarrow{EF} = \overrightarrow{AB} + (\overrightarrow{CD} + \overrightarrow{EF})$.

THEOREM V.2.2. *Vector addition is commutative.*

Proof. Let $\overrightarrow{AB}$, $\overrightarrow{CD}$ be any two vectors, and let $\overrightarrow{BQ} = \overrightarrow{CD}$. Then $\overrightarrow{AB} + \overrightarrow{CD} = \overrightarrow{AQ}$. Suppose $\overrightarrow{AB}$, $\overrightarrow{BQ}$ have distinct carriers, and let the line on A, parallel to line BQ, meet the line on Q, parallel to line AB, at P. Then $\overrightarrow{CD} = \overrightarrow{BQ} = \overrightarrow{AP}$, and $\overrightarrow{AB} = \overrightarrow{PQ}$ yields $\overrightarrow{CD} + \overrightarrow{AB} = \overrightarrow{AP} + \overrightarrow{PQ} = \overrightarrow{AQ}$.

If, on the other hand, $\overrightarrow{AB}$, $\overrightarrow{BQ}$ have the same carrier, let S be a point of Π not on line AB (Postulate 3, Section IV.1). Then
$$\overrightarrow{CD} + \overrightarrow{AB} = \overrightarrow{BQ} + \overrightarrow{AB} = (\overrightarrow{BS} + \overrightarrow{SQ}) + \overrightarrow{AB}$$
$$= \overrightarrow{BS} + (\overrightarrow{SQ} + \overrightarrow{AB}) \quad \text{(Theorem V.2.1)}$$
$$= \overrightarrow{BS} + (\overrightarrow{AB} + \overrightarrow{SQ}) \quad \text{(since } \overrightarrow{AB}, \overrightarrow{SQ} \text{ have different}$$
$$\text{carriers)}$$
$$= (\overrightarrow{BS} + \overrightarrow{AB}) + \overrightarrow{SQ}$$
$$= (\overrightarrow{AB} + \overrightarrow{BS}) + \overrightarrow{SQ} = \overrightarrow{AS} + \overrightarrow{SQ} = \overrightarrow{AQ},$$

and the proof is complete.

These two theorems have important consequences in regard to the algebraic structure of the ternary ring $[\Gamma, T]$. A one-to-one correspondence between the set of vectors $\overrightarrow{OX}$ and the points X of the unit line is obtained by associating with each such vector its terminal point X (the origin O is associated with the null vector $\overrightarrow{OO}$). This one-to-one correspondence, together with the one-to-one correspondence γ between the points of the unit line and the elements of the coordinate set Γ (see Section IV.3) establishes a one-to-one correspondence between the vectors $\overrightarrow{OX}$, $X \in$ line OI, and the elements of Γ, in which $\overrightarrow{OA}$ corresponds to a, where $a = \gamma(A) \in \Gamma$.

If $A(a, a)$, $B(b, b) \in$ line OI, and $C = A + B$ (see Figure 11), $\overrightarrow{OA} = \overrightarrow{UP} = \overrightarrow{BC}$, so $\overrightarrow{OA} + \overrightarrow{OB} = \overrightarrow{BC} + \overrightarrow{OB} = \overrightarrow{OB} + \overrightarrow{BC} = \overrightarrow{OC}$. Hence if A, $B \in$ line OI that correspond, respectively, to vectors $\overrightarrow{OA}$, $\overrightarrow{OB}$, their sum $A + B$ corresponds to the vector sum $\overrightarrow{OA} + \overrightarrow{OB}$, and, consequently, if a, $b \in \Gamma$ that correspond, respectively, to points A, B (and hence to vectors $\overrightarrow{OA}$, $\overrightarrow{OB}$), their sum $a + b$ corresponds to $C = A + B$, and hence to $\overrightarrow{OC} = \overrightarrow{OA} + \overrightarrow{OB}$. Thus, the vector that corresponds to the sum of a and b is the sum of the vector corresponding to a and the vector corresponding to b. We conclude that this correspondence between vectors $\overrightarrow{OX}$, $X \in$ line OI, and elements x of Γ is an *isomorphism with respect to addition*. Since vector addition is associative and commutative, it follows that *addition in* $[\Gamma, T]$ *is associative and commutative also*. Suppose a, $b \in \Gamma$ and $a \sim \overrightarrow{OA}$, $b \sim \overrightarrow{OB}$ in the correspondence defined above (where the symbol $\sim$ is read "corresponds to"). Then $a + b \sim \overrightarrow{OA} + \overrightarrow{OB}$, and $b + a \sim \overrightarrow{OB} + \overrightarrow{OA}$, and, since $\overrightarrow{OA} + \overrightarrow{OB}$ and $\overrightarrow{OB} + \overrightarrow{OA}$ are the *same* vector $\overrightarrow{OC}$, and the correspondence is one-to-one, $a + b = b + a$. In a similar manner addition in $[\Gamma, T]$ is shown to be associative.

We have proved that with respect to ordered addition (Remark 2, Section IV.8) $[\Gamma, T]$ is a loop. Imposing the first Desargues property on plane Π results in associativity of addition in $[\Gamma, T]$, hence $[\Gamma, +]$ is a *group* (a loop with associativity of the operation). Since addition is, moreover, commutative, the group $[\Gamma, +]$ is an *abelian* (that is, commutative) group.

We have now established the following important result.

THEOREM V.2.3.   *If a rudimentary affine plane Π has the first Desargues property, every planar ternary ring $[\Gamma, T]$ defined over Π is an abelian group with respect to addition.*

V.3. Linearity of the Ternary Operator

Without assuming the affine plane Π to have the first Desargues property, we have seen that certain classes of lines of Π are represented by equations having the same forms as in ordinary analytic geometry. Thus lines in the parallel class of the y-line have equations $x = a$, $a \in \Gamma$; those in the parallel class of the x-line have equations $y = b$, $b \in \Gamma$; those in the parallel class of the unit line have equations $y = x + c$, $c \in \Gamma$; and those lines on O with slope m, $m \in \Gamma$, have equations $y = xm$ (where the dot indicating multiplication has been omitted). Now it will be established that, as a consequence of assuming the first Desargues property valid in Π, each line of Π with slope m and y-intercept b has an equation $y = xm + b$, as in ordinary analytic geometry.

Consider the line with equation $y = T(x, m, b)$, where $m \neq 0, 1$ and $b \neq 0$ (Figure 18).

The two triples A^*, A', A'' and B^*, B', B'' of Figure 18 are such that: (1) the six points are pairwise distinct, (2) lines A^*B^*,

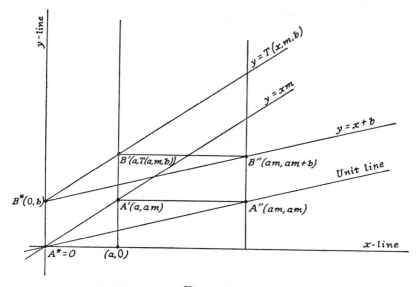

Figure 18

$A'B'$, $A''B''$ are pairwise mutually parallel, and (3) line $A*A'$ is parallel to line $B*B'$, and line $A*A''$ is parallel to line $B*B''$. It follows from the first Desargues property that line $A'A''$ is parallel to line $B'B''$, and hence line $B'B''$ is in the parallel class of the x-line. Consequently, points B', B'' have the same ordinates, so $T(a, m, b) = am + b$, $a \neq 0$. By Section IV.6, $T(0, m, b) = b = T(a, 0, b)$; by Section IV.8, $T(a, 1, b) = a + b$; and by Section IV.9, $T(a, m, 0) = am$. Hence for every triple a, m, $b \in \Gamma$, $T(a, m, b) = am + b$, and each line with slope m and y-intercept b has equation $y = T(x, m, b) = xm + b$. Thus $T(x, m, b)$ is linear.

Since $x \cdot y = T(x, y, 0)$, $t + z = T(t, 1, z)$ $(x, y, z, t \in \Gamma)$, it follows that $T(x, y, z) = xy + z = T(xy, 1, z) = T[T(x, y, 0), 1, z]$.

We recall that a coordinate system was established in the affine plane Π by selecting an arbitrary point O of Π for the origin and by choosing three pairwise distinct lines on O and designating them the x-line, the y-line, and the unit line. On the unit line any point distinct from O was chosen and labeled I, and an arbitrary abstract set Γ was selected (subject to the sole requirement that there is a one-to-one correspondence γ between the points of the line OI and the elements of Γ), and $\gamma(O)$, $\gamma(I)$ were labeled 0, 1, respectively. Then a procedure was defined that established a one-to-one correspondence between the points of Π and the elements of the set of all ordered pairs of elements of Γ. The ordered pair of elements of Γ corresponding to a point P of Π are the coordinates of P. A ternary operation $T(a, m, b)$ was defined in Γ, and the system $[\Gamma, T]$ was called a *planar ternary ring*.

Now if a different point O' of Π were chosen for the origin, and three lines on O' selected as the x'-line, y'-line, and unit line (with point I' selected on it), the same set Γ could serve as the coordinate set, since there exists a one-to-one correspondence between the elements of Γ and the points of the new unit line [each line of Π contains the same number (finite or transfinite) of points]. But the ternary operation T', defined by the new lines and new origin, would, in general, be quite different from the ternary operation T defined by the original coordinate system (though the *domain* of the two operators is the same set Γ), so a different planar ternary ring $[\Gamma, T']$ would result. If the first Desargues property is assumed for Π, both operators T, T' are linear; that is $T(x, y, z) = xy + z$, $T'(x, y, z) = [x \odot y] \oplus z$, where the addition $\oplus$ and the multiplication $\odot$ are, in general, different from the addition $+$ and multiplication $\cdot$ defined by using the original coordinate system. We have established the following result.

THEOREM V.3.1. *If the first Desargues property is valid in the affine plane Π, every (planar) ternary ring defined in Π is linear.*

Is the converse of this theorem valid; that is, if every (planar) ternary ring defined in the affine plane Π is linear, does it follow that the first Desargues property is valid in Π? The answer is affirmative, as we shall now show.

THEOREM V.3.2. *If every (planar) ternary ring defined in an affine plane Π is linear, Π has the first Desargues property.*

Proof. Let A^*, A', A'' and B^*, B', B'' be any two triples of points of Π (each two points distinct) such that lines A^*B^*, $A'B'$, $A''B''$ are pairwise mutually parallel and lines A^*A', $A'A''$ are parallel, respectively, to lines B^*B', $B'B''$. We wish to prove that line A^*A'' is parallel to line B^*B''.

We introduce in Π a coordinate system in the following manner. Select A' as the origin, line $A'B'$ as the y-line, the line on A' parallel to line A^*A'' as the x-line, and line $A'A''$ as the unit line (Figure 19).

Then line $A'A^*$ has equation $y = xm$, and line $B'B^*$ has equation $y = T(x, m, b)$ for some element b of Γ (since it is parallel to line $A'A^*$). Since line $B'B''$ is parallel to the unit line, it has equation $y = x + b$. Point A^* has coordinates (a, am) for some element a of Γ, so point A'' has coordinates (am, am), and the coordinates of B^* and B'' are $(a, T(a, m, b))$ and $(am, am + b)$, respectively. But since every ternary ring defined in Π is linear, $T(a, m, b) = am + b$, and, consequently, points B^* and B'' have

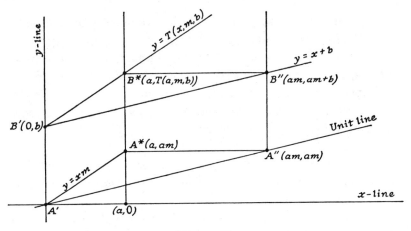

Figure 19

the same ordinates. It follows that line B^*B'' is in the parallel class of the x-line, and since line A^*A'' is also in that parallel class, and is distinct from line B^*B'', it follows that line A^*A'' is parallel to line B^*B''.

Combining the last two theorems yields the following important result.

THEOREM V.3.3.   *The first Desargues property is valid in an affine plane* Π *if and only if every (planar) ternary ring defined in* Π *is linear.*

The (geometric) first Desargues property has, then, the algebraic counterpart or *equivalent* that every ternary ring of Π is linear.

V.4. Right Distributivity of Multiplication Over Addition

We have obtained two important consequences of assuming an affine plane Π has the first Desargues property: *the system* $[\Gamma, +]$ *is an abelian group, and the ternary operator T is linear.* A third important consequence of that assumption is established in the following theorem.

THEOREM V.4.1. *If an affine plane* Π *has the first Desargues property, multiplication is right distributive over addition [that is, for any three points A, B, P of the unit line, $(A + B) \cdot P = A \cdot P + B \cdot P$].*

Proof. It may be assumed that each of the points A, B, P is distinct from the origin O, since otherwise the property to be proved is obvious.

Let (a, a), (b, b), (p, p) be the coordinates of points A, B, P, respectively, and consider the points $B^*(0, b)$, $C(a + b, a + b)$, $C^*(a, a + b)$ (Figure 20).

Line B^*C^* has equation $y = x + b$; the line on O with slope p has equation $y = xp$ and meets the line on B, in the parallel class of the y-line, in the point $A'(b, bp)$. The line on A', in the parallel class of the x-line, meets the y-line in point $A^*(0, bp)$. The line on A^* with slope p has equation $y = xp + bp$, and this line meets line AC^* in point $D(a, ap + bp)$. Finally, the line on D, in the

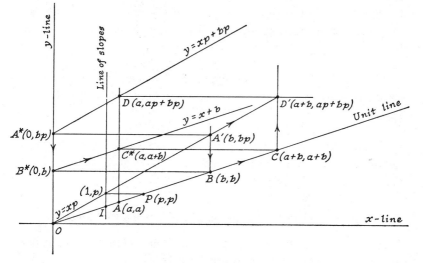

Figure 20

parallel class of the x-line, meets the line on C, in the parallel class of the y-line, in point $D'(a + b, ap + bp)$.

It suffices to show that D' is on line OA' (which has equation $y = xp$), for then $ap + bp$ and $(a + b)p$ are both ordinates of D', and, consequently, are equal. To this end we consider the vector $\overrightarrow{A'D'} = \overrightarrow{A'B} + \overrightarrow{BC} + \overrightarrow{CD'}$. Now $\overrightarrow{A'B} = \overrightarrow{A^*B^*}$, $\overrightarrow{BC} = \overrightarrow{B^*C^*}$, and $\overrightarrow{CD'} = \overrightarrow{C^*D}$. Hence $\overrightarrow{A'D'} = \overrightarrow{A^*B^*} + \overrightarrow{B^*C^*} + \overrightarrow{C^*D} = \overrightarrow{A^*D}$, so line $A'D'$ is parallel to line A^*D. But line OA' is parallel to line A^*D, so by Postulate 2 of an affine plane Π (Section IV.1) lines OA' and $A'D'$ are identical. Hence point D' is on line OA', and the theorem is proved.

Remark 1.   In Figure 20 the points A, B, P are selected pairwise distinct. Let the reader verify that the same argument may be used in the event any two of the points (or even all three of them) coincide.

Remark 2.   Each planar ternary ring $[\Gamma, T]$ of an affine plane Π with the first Desargues property has the following properties: (1) $[\Gamma, +]$ is an abelian group with identity element 0; (2) $[\Gamma', \cdot]$ is a loop with identity 1; (3) for every element a of Γ, $a \cdot 0 = 0 = 0 \cdot a$; (4) for all a, b, $c \in \Gamma$, $(a + b) \cdot c = ac + bc$. Such an algebraic structure is called a *Veblen-Wedderburn system*, and we have obtained the following result.

THEOREM V.4.2.   *If an affine plane Π has the first Desargues property, every planar ternary ring of Π is a Veblen-Wedderburn system.*

It may be shown that if *any one* of the planar ternary rings defined in an affine plane Π is a Veblen-Wedderburn system, Π has the first Desargues property.

V.5. Introduction of the Second Desargues Property

Imposing the first Desargues property on the affine plane Π has advanced us considerably toward our objective of determining those properties of the plane that permit its coordinatization by real numbers. An example shows that *left* distributivity of multiplication over addition cannot be proved on the basis of our present assumptions, and, since that property of $[\Gamma, T]$ is desired, we introduce another assumption that will secure it.

THE SECOND DESARGUES PROPERTY. *Let seven pairwise distinct points of Π consist of the two triples A, B, C and A', B', C' (each triple being non-collinear) and the point O, collinear with each of the point pairs (A, A'), (B, B'), (C, C'). If line AB is parallel to line $A'B'$ and lines $BC, B'C', OA$ are pairwise mutually parallel, line AC is parallel to line $A'C'$. (Figure 21).*

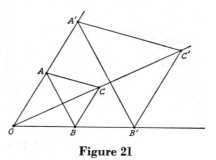

Figure 21

THEOREM V.5.1. *If an affine plane Π has the second Desargues property, in each planar ternary ring $[\Gamma, T]$ of Π, $T(a, b, ac) = a \cdot T(1, b, c)$, for all elements a, b, c of Γ.*

Proof. Since the assertion follows from previous results (Sections IV.6 and IV.9), in the event $a = 0$ or $b = 0$ or $c = 0$, we may suppose that no one of these elements of Γ is 0.

Consider the points $(1, 0)$, $(a, 0)$, $B(1, c)$, and the line OB, which has equation $y = xc$ (Figure 22).

The line on B, parallel to the x-line, meets the y-line in $A(0, c)$. Let C denote the point with coordinates $(1, T(1, b, c))$. Then line OC has equation $y = x \cdot T(1, b, c)$, and the line on $(a, 0)$, parallel to the y-line, meets line OB in $B'(a, ac)$ and line OC in $C'(a, a \cdot T(1, b, c))$.

Now line AC has equation $y = T(x, b^*, c)$, for some element b^* of Γ. We show that $b^* = b$.

If $b^* \neq b$, the equation $T(x, b^*, c) = T(x, b, c)$ has a *unique* solution (Section IV.6). Since $T(0, b^*, c) = c = T(0, b, c)$, $x = 0$ is one solution of the equation. But since C is on line AC, its

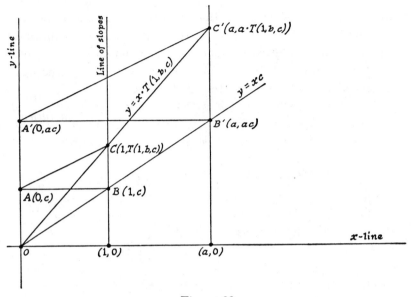

Figure 22

coordinates $(1, T(\mathbf{1}, b, c))$ satisfy the equation $y = T(x, b^*, c)$ of that line, so $T(1, b, c) = T(1, b^*, c)$. Hence $x = 1$ is also a solution of the equation $T(x, b^*, c) = T(x, b, c)$, and since $0 \neq 1$, the uniqueness of the solution of the equation is violated. This contradiction establishes that $b^* = b$, so line AC has equation $y = T(x, b, c)$, and, consequently, has slope b. Let $A'(0, ac)$ denote the intersection of the y-line with the line on B' parallel to the x-line.

The seven points O, A, B, C, A', B', C' satisfy the hypothesis of the second Desargues property, hence line AC is parallel to line $A'C'$. It follows that the slope of line $A'C'$ is b, and that line has an equation $y = T(x, b, ac)$. Since C' is a point of line $A'C'$, and since C' has coordinates $(a, a \cdot T(1, b, c))$, $a \cdot T(1, b, c) = T(a, b, ac)$, and the theorem is proved.

The reader will observe it is not assumed in the preceding theorem that Π has the first Desargues property, and, consequently, the algebraic equivalent of that property—the linearity of the ternary operator T—is not available.

COROLLARY V.5.1.　*If an affine plane* П *has the first and second Desargues properties, then, for all elements a, b, c of any planar ternary ring* [Γ, T] *of* П, $a(b + c) = ab + ac$ *(that is, multiplication in* [Γ, T] *is left distributive over addition).*

Proof. By the preceding theorem, $T(a, b, ac) = a \cdot T(1, b, c)$ for all $a, b, c \in \Gamma$, and, according to Theorem V.3.3, the ternary operator T is linear. Hence $ab + ac = a(1 \cdot b + c) = a(b + c)$.

Summarizing, when both the first and second Desargues properties are valid in an affine plane, every planar ternary ring [Γ, T] of П has the following properties:

(1) [Γ, $+$] is an abelian group, with identity 0.

(2) [Γ′, $\cdot$] is a loop, with identity 1.

(3) For every element a of Γ, $a \cdot 0 = 0 = 0 \cdot a$.

(4) For all elements a, b, c of Γ, $(a + b)c = ac + bc$ and $c(a + b) = ca + cb$.

Such an algebraic structure is called by some writers a *division ring*.

V.6. Introduction of the Third Desargues Property

We turn our attention now to the associativity of multiplication, and to secure it we introduce the third Desargues property.

THE THIRD DESARGUES PROPERTY.　*Let seven pairwise distinct points of an affine plane* П *consist of two triples A, B, C and A′, B′, C′ (each triple being non-collinear) and the point O, collinear with each of the point pairs* (A, A'), (B, B'), (C, C'). *If lines AB and A′B′ are mutually parallel, and lines BC and B′C′ are mutually parallel, either line AC coincides with line A′C′ or the two lines are mutually parallel.*

The reader will note that *the third Desargues property implies the second Desargues property,* since it asserts the same result without the restriction that lines BC and $B'C'$ be parallel to line OA, which is required by the second Desargues property.

THEOREM V.6.1.   *If an affine plane* Π *has the third Desargues property, then, in any ternary ring* $[\Gamma, T]$ *defined in* Π, *multiplication is associative.*

Proof.   Let $A(a, a)$, $B(b, b)$, $C(c, c)$ be points of the unit line OI. We wish to prove that $a(bc) = (ab)c$. Since this relation is clearly satisfied in the event any one of the three elements a, b, c of Γ is 0 or 1 (Remarks 1 and 6, Section IV.9), we suppose this is not the case; that is, each of the points A, B, C is assumed different from O or I.

Case 1.   *The points* A, B, C *are pairwise distinct.* The line on O and point $(1, c)$ has equation $y = xc$ and intersects line $x = b$ in point $R(b, bc)$ (Figure 23). The line on points O and $P(1, b)$ has equation $y = xb$ and meets line $x = a$ in point $P'(a, ab)$. Similarly, the line on O and point $Q(1, bc)$ has equation $y = x(bc)$ and meets line $x = a$ in $Q'(a, a(bc))$. Line $y = ab$ is on point $P'(a, ab)$

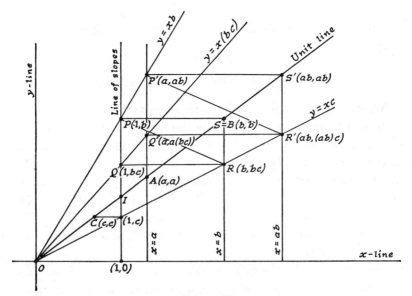

Figure 23

and meets the unit line in point $S'(ab, ab)$, and line $x = ab$ is on point S' and meets line $y = xc$ in point $R'(ab, (ab)c)$. Denote point B on the unit line also by S.

Consideration of the triples P, R, S and P', R', S' shows that the point pairs (P, P'), (R, R'), (S, S') are collinear with O, line PS is parallel to line $P'S'$, and line RS is parallel to line $R'S'$. Hence, by the third Desargues property, line PR is parallel to line $P'R'$.

Now the triples P, Q, R, and P', Q', R' are such that each of the pairs (P, P'), (Q, Q'), (R, R') is collinear with point O, line PQ is parallel to line $P'Q'$, and (as just shown) line PR is parallel to line $P'R'$. Hence lines QR and $Q'R'$ are mutually parallel, and since line QR is parallel to the x-line, so is line $Q'R'$, and points Q', R' have the same ordinate. Consequently, $a(bc) = (ab)c$, and the theorem is proved in this case.

Case 2. $A = B \neq C$ or $A = C \neq B$. The argument in Case 1 may be applied without change to prove the theorem in this case.

Case 3. $B = C \neq A$. According to Case 2, if a, $b \in \Gamma$, $a \neq b$, then $a(ab) = (aa)b$ and $a(ba) = (ab)a$. We wish to show $a(bb) = (ab)b$.

The procedure of Case 1, applied here, gives Figure 24.

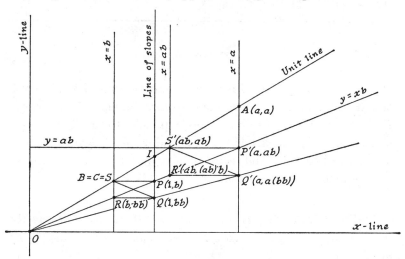

Figure 24

Applying the third Desargues property to the triples P, Q, S and P', Q', S' gives line QS parallel to line $Q'S'$, and applying it to triples Q, R, S and Q', R', S' yields line QR parallel to line $Q'R'$. It follows that Q' and R' have the same ordinates, and, consequently, $a(bb) = (ab)b$.

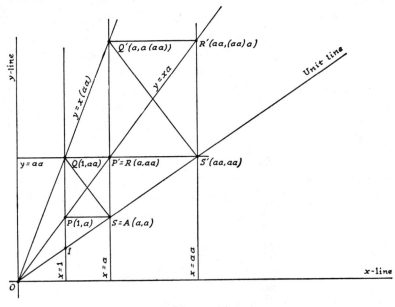

Figure 25

Case 4. $A = B = C$. Employing again the procedure of Case 1 (Figure 25), the third Desargues property applied to the triples P, Q, S and P', Q', S' gives line QS parallel to line $Q'S'$. Consideration of the two triples Q, R, S and Q', R', S' yields line QR parallel to line $Q'R'$, and, consequently, $a(aa) = (aa)a$, completing the proof of the theorem.

. THEOREM V.6.2.   *If an affine plane* Π *has the first and the third Desargues properties, then, for every planar ternary ring* $[\Gamma, T]$ *defined in* Π:

(1) $[\Gamma, +]$ *is an abelian group, with identity* **0**.

(2) $[\Gamma', \cdot]$ *is a group, with identity* **1**.

(3) *For every element a of Γ, $a \cdot 0 = 0 = 0 \cdot a$.*

(4) *For all elements a, b, c of Γ, $(a + b)c = ac + bc$ and $c(a + b) = ca + cb$.*

Proof. Since multiplication in Γ is *associative*, the loop $[\Gamma', \cdot]$ is a group.

Remark. An algebraic structure with the four properties listed in Theorem V.6.2 is called a *field* or *skew field*, depending on whether or not multiplication is *commutative*.

V.7. Introduction of the Pappus Property

We seek now a geometrical property to impose on an affine plane Π that will be the logical equivalent of the algebraic property of commutativity of the multiplication defined (Section IV.9) in the planar ternary ring $[\Gamma, T]$ of Π. It is found in a theorem of classical geometry established by the Greek mathematician, Pappus, who was born in Alexandria about 350 A.D. and who was probably the last great figure of that famous school of mathematics.

THE PAPPUS PROPERTY. *Let P, Q, R be points of one line and P', Q', R' be points of another line (all six points being pairwise distinct, and none being on both of the lines), and let lines PQ' and $P'Q$ be mutually parallel or intersect in a point S, and let lines QR' and $Q'R$ be mutually parallel or intersect in a point T. If the first alternative holds in each case, lines PR' and $P'R$ are mutually parallel, whereas if the second alternative holds in each case, line ST contains the point in which lines PR' and $P'R$ meet, or the lines PR', $P'R$, ST are parallel to one another.*

If the first alternative holds, the property is called the *affine Pappus property* (Figure 26).

We consider the situation illustrated in Figure 27 as a *special case* of the Pappus property stated above; that is, if S is the point in which line $P'Q$ meets the line on Q' that is parallel to line QR, and T is the point in which the line $P'R$ meets the line on R' that is parallel to line QR, $S \neq T$, and line QR' is parallel to line $Q'R$, then line ST is parallel to line QR'. Let us call this the *special Pappus property*. Since $S \neq T$, line QR is not parallel to line $P'Q'$.

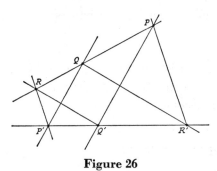

Figure 26

THEOREM V.7.1. *Multiplication in every planar ternary ring* $[\Gamma,\ T]$ *of an affine plane* Π *is commutative if and only if* Π *has the special Pappus property.*

Proof. Suppose, first, that Π has the special Pappus property, and let $A(a, a)$, $B(b, b)$ be two points of the unit line in an arbitrarily selected coordinate system of Π. Since commutativity of multiplication is obvious, or follows at once from its definition, in the event $A = B$, or if either point is O or I, we consider $A \neq B$, and neither point O nor point I.

Constructing the products AB and BA (Figure 28) we observe that the points Q, R and O, A, B satisfy the conditions of the special case of the Pappus property described in the paragraph preceding this theorem. It follows that line ST is parallel

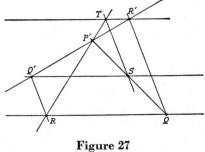

Figure 27

to line QB, and, consequently, S and T have the same ordinates; that is $ab = ba$.

Conversely, let multiplication be assumed commutative in every planar ternary ring of Π, and let Y, Z and X', Y', Z' be any five

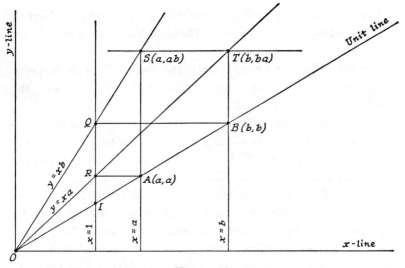

Figure 28

points satisfying the hypothesis of the special Pappus property. Choose a coordinate system in Π such that X' is the origin, the line containing X', Y', Z' is the unit line, the y-line is the line on X', parallel to line YZ, the x-line is the line on X', parallel to line $Y'Z$ (the labeling of the five points has been selected such that lines $Y'Z$ and YZ' are mutually parallel), and I is the intersection of lines YZ and $X'Y'$. Since multiplication in the planar ternary ring associated with this coordinate system is commutative, and since the points S, T have coordinates $(y', y' \cdot z')$, $(z', z' \cdot y')$ in this system [where (y', y') and (z', z') are the coordinates of Y' and Z', respectively], $y' \cdot z' = z' \cdot y'$, so line ST is parallel to line YZ'. Hence the special Pappus property is valid in Π.

Let the reader construct the figure.

COROLLARY V.7.1. *If an affine plane Π has the first and third Desargues properties and the Pappus property, every planar ternary ring of Π is a field.*

Commutativity of multiplication is all that need be added to the properties of $[\Gamma, T]$ listed in Theorem V.6.2 to make $[\Gamma, T]$ a field.

Remark. Since we needed only the special Pappus property to obtain commutativity of multiplication, the hypothesis of the corollary may be weakened.

We have now obtained a major result of our study of the problem of coordinatizing the affine plane. Starting with a completely abstract "coordinate set" Γ, we defined geometrically a ternary operation T in Γ to obtain the planar ternary ring $[\Gamma, T]$, and properties of this ring were explored. Geometrically (or by use of the operator T), addition and multiplication were defined in $[\Gamma, T]$, and geometric properties were successively imposed on the plane Π that resulted in $[\Gamma, T]$ becoming, in turn, a Veblen-Wedderburn system, a division ring, and now a field. We have come a long way toward accomplishing the coordinatization of a plane in the manner of elementary analytic geometry (that is, a one-to-one correspondence between the set of all points of the plane and the set of all ordered pairs of *real numbers*).

Our next task is to show that the conclusion of the above corollary may be retained even when the requirement is deleted that Π have the first and third Desargues properties!

V.8. The Desargues Properties as Consequences of the Pappus Property

We begin the elimination of the Desargues properties, as *explicit* requirements, with the following theorem, which the German mathematician, Gerhard Hessenberg, proved in 1905.

THEOREM V.8.1. *In an affine plane Π the affine Pappus property implies the third Desargues property.*

Proof. Let A, B, C and A', B', C' be two non-collinear triples such that lines AA', BB', CC' are on the point O and the seven points are pairwise distinct. (See Figure 29.) Assuming that lines

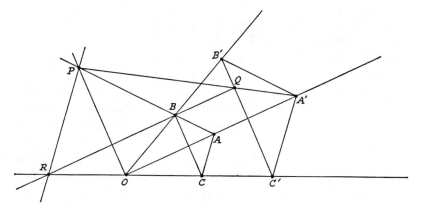

Figure 29

AB, A'B' are mutually parallel, that line *BC* is parallel to line *B'C'*, and that line *AC* ≠ line *A'C'*, we show that line *AC* is parallel to line *A'C'*.

The line on *O*, parallel to line *BC*, meets line *AB* at a point *P* (otherwise the distinct lines *AB* and *BC* are each parallel to that line, in violation of Postulate 2, Section IV.1). Lines *A'P* and *B'C'* meet at some point *Q* (if they do not, each of the distinct lines *A'P* and *OP* is parallel to line *B'C'*).

We assert that lines BQ and OC intersect; consider the two sets of collinear triples *A', P, Q* and *B, B', O*. Lines *A'B'* and *BP* are mutually parallel, and line *OP* is parallel to line *B'Q*. Then, by the affine Pappus property, lines *OA'* and *BQ* are mutually parallel. Consequently, lines *BQ* and *OC* intersect, otherwise the distinct lines *OA'* and *OC* are each parallel to *BQ*. Let *R* denote the intersection of lines *BQ* and *OC*.

Examining the two sets of collinear triples *R, O, C* and *A, B, P* we find that lines *AO* and *BR* are mutually parallel, as are lines *BC* and *OP*. Invoking the affine Pappus property again we conclude that line *AC* is parallel to line *PR*.

Finally, consideration of the two collinear triples *P, Q, A'* and *C', O, R* yields line *OP* parallel to line *C'Q* and line *QR* parallel to line *OA'*, and it follows from the affine Pappus property that line $A'C'$ is parallel to line PR. Hence each of the lines AC and $A'C'$ is parallel to line PR, and since line $AC \neq$ line $A'C'$, we conclude that line AC is parallel to line $A'C'$.

We remark that the points in Figure 29 have been labeled so that line OAA' is not parallel to lines BC and $B'C'$. If such a labeling is not possible, a slight modification of the proof is needed.

THEOREM V.8.2. *Let* Π *be an affine plane with the third Desargues property. If the corresponding lines AB and $A'B'$, BC and $B'C'$, AC and $A'C'$ of two point triples A, B, C and A', B', $C'*

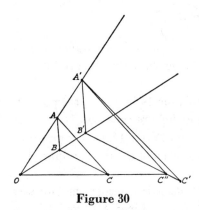

Figure 30

(each non-collinear) are mutually parallel, the three lines AA', BB', CC' are pairwise mutually parallel or are on a common point.

Proof. Suppose the lines AA' and BB' meet at a point O (Figure 30). Line OC intersects line $B'C'$ in a point C'' (otherwise the distinct lines OC and BC would each be parallel to line $B'C'$).

Examining the two sets of non-collinear triples A, B, C and A', B', C'' we observe that lines AB and $A'B'$ are mutually parallel, as are lines BC and $B'C''$, and each of the point pairs (A, A'), (B, B'), (C, C'') is collinear with O. Applying the third Desargues property, it follows that lines AC and $A'C''$ are mutually parallel. Since, by hypothesis, line AC is parallel to line $A'C'$, lines $A'C'$ and $A'C''$ are coincident, and $C'' \in$ line $A'C'$. But C'' is also a point of line $B'C'$, and since lines $A'C'$ and $B'C'$ are distinct, it follows that $C'' = C'$. Hence O, C, C' are collinear, and the three lines AA', BB', CC' are on the common point O.

If, on the other hand, lines AA' and BB' are mutually parallel, line CC' must be parallel to each of those lines, since if it intersected either of them, the other would contain that intersection point (by the first part of the proof), and the lines AA', BB' would not be mutually parallel.

THEOREM V.8.3. *If an affine plane Π has the Pappus property, it has the first Desargues property.*

Proof. Let six pairwise distinct points of Π form two triples A, B, C, and A', B', C' such that lines AA', BB', CC' are pairwise mutually parallel, line AB is parallel to line $A'B'$, and line BC is parallel to line $B'C'$ (Figure 31). We wish to show that line AC is parallel to line $A'C'$. This is obvious in the event A, B, C are collinear, so we assume they are not.

The line on A', parallel to line AC, intersects the line $B'C'$ in a point C'' (why?). Examin-
ing the triples A, B, C and A', B', C'' we see that line AB is parallel to line $A'B'$, line BC is parallel to line $B'C''$, and line AC is parallel to line $A'C''$. It follows from the two preceding theorems

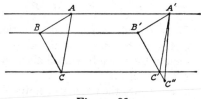

Figure 31

of this section that the lines AA', BB', CC'' are either pairwise mutually parallel or are on a common point. But lines AA' and BB' are mutually parallel, hence so are lines BB' and CC''. Since line CC' is parallel to line BB', lines CC' and CC'' coincide, and clearly $C' = C''$. Then line $A'C'$ is parallel to line AC, and the theorem is proved.

Combining Theorems V.8.1 and V.8.3 with Corollary V.7.1 yields the following important result.

THEOREM V.8.4. *If an affine plane Π has the Pappus property, every planar ternary ring of Π is a field.*

Thus the points of any affine plane Π with the Pappus property are in a one-to-one correspondence with the set of all ordered pairs of elements of a field $\mathfrak{F}$. The geometry of the plane is called an *affine geometry over the field* $\mathfrak{F}$.

● **EXERCISES**

1. An affine plane Π is formed by the set of 25 marks (*points*) $A, B, C, D, \cdots, U, V, W, X, Y$ (all of the letters of the English alphabet except Z) when any five letters that occur together in any row or any column of the following three blocks are defined to constitute a *line*.

	(1)					(2)					(3)			
A	B	C	D	E	A	I	L	T	W	A	X	Q	O	H
F	G	H	I	J	S	V	E	H	K	R	K	I	B	Y
K	L	M	N	O	G	O	R	U	D	J	C	U	S	L
P	Q	R	S	T	Y	C	F	N	Q	V	T	M	F	D
U	V	W	X	Y	M	P	X	B	J	N	G	E	W	P

Check Postulates 1, 2, and 3, Section IV. 1.

2. Let Γ consist of the marks 0, 1, 2, 3, 4. Establish a coordinate system in the plane Π defined above with coordinates being ordered pairs of elements of Γ.

3. Show how the ternary operator T is defined, finding $T(a, m, k)$ for all of the ordered triples of pairwise distinct elements of the quadruple 1, 2, 3, 4. (Construct a table.)

4. Obtain the "value" of $a \cdot m$ and of $a \cdot m + k$ for all of the ordered triples used in Exercise 3. Compare $T(a, m, k)$, from Exercise 3, with $am + k$.

5. Examine Π for Desargues and Pappus properties.

6. The theorems of Sections V.7 and V.8 have been obtained by using only the affine Pappus and special Pappus properties. Does either of these properties imply the other in an affine plane Π?

7. How many coordinate systems are possible in the above plane?

V.9. Analytic Affine Geometry Over a Field

According to Theorem V.8.4, the geometry of an affine plane Π, with the Pappus property, may be developed analytically (algebraically) in much the same way as euclidean geometry is treated in the usual course of analytic geometry. Points are ordered pairs (x, y) of elements of a field $\mathfrak{F}$, and lines are linear equations

$$a_1 x + a_2 y + a_3 = 0, \quad (a_1, a_2, a_3 \in \mathfrak{F}, \quad a_1, a_2 \neq 0, 0).$$

By an *affine property* is meant any logical consequence of Postulates 1, 2, and 3 (Section IV.1) and the Pappus property. Since such a property is clearly independent of the choice of a coordinate system, the analytic expression of that property must be invariant under a change (transformation) of coordinates. How is a change of coordinate systems expressed analytically?

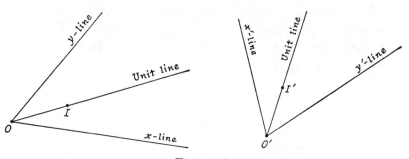

Figure 32

Suppose it is desired to change from the (x, y)-coordinate system of Figure 32 to the (x', y')-coordinate system. Let $a_1x + a_2y + a_3 = 0$ and $b_1x + b_2y + b_3 = 0$ be the equations of the y'-line and x'-line, respectively, in the (x, y)-coordinate system, and let (c, d) be the coordinates of the point I' in that system. Then the equations

$$x' = \alpha \cdot (a_1x + a_2y + a_3),$$
$$y' = \beta \cdot (b_1x + b_2y + b_3),$$

where $\alpha = (a_1c + a_2d + a_3)^{-1}$, $\beta = (b_1c + b_2d + b_3)^{-1}$ give the coordinates (x', y') in the "new" system of any point P with coordinates (x, y) in the "old" system. (If $g \in \mathfrak{F}, g \neq 0, g^{-1}$ denotes the inverse of g with respect to *multiplication;* that is, g^{-1} is the unique element of $\mathfrak{F}$ such that $g \cdot g^{-1} = g^{-1} \cdot g = 1$). Note that since point I' is not on either of the lines $a_1x + a_2y + a_3 = 0$, $b_1x + b_2y + b_3 = 0$, then $a_1c + a_2d + a_3 \neq 0$, $b_1c + b_2d + b_3 \neq 0$, and the unique respective inverses α, β of those elements exist. Each point (x, y) on the line $a_1x + a_2y + a_3 = 0$ has new coordinates $(0, y')$ and hence is on the y'-line, and each point (x, y) on the line $b_1x + b_2y + b_3 = 0$ has new coordinates $(x', 0)$ and hence is on the x'-line, and the new coordinates of I' are obviously $(1, 1)$. Since the x'-line and the y'-line are neither coincident nor mutually parallel, $a_1b_2 - a_2b_1 \neq 0$. [If $g \in \mathfrak{F}$, $-g$ denotes the inverse of g with respect to *addition;* that is $-g$ symbolizes that unique element of $\mathfrak{F}$ such that $g + (-g) = 0 = (-g) + g$.]

The expression $x_1 - x_2$ does not represent an affine property of points $P_1(x_1, y_1)$, $P_2(x_2, y_2)$, since its value clearly depends on the coordinate system to which the points P_1, P_2 are referred. On the other hand, the *vanishing* of the determinant

$$\begin{vmatrix} x_1 & y_1 & 1 \\ x_2 & y_2 & 1 \\ x_3 & y_3 & 1 \end{vmatrix}$$

is an affine property of the points $P_i(x_i, y_i)$ $(i = 1, 2, 3)$ (though the *value* of the determinant is not, if it is different from zero). Let the reader verify this and ascertain what affine property of the three points is expressed by the vanishing of the determinant.

Thus, a change of coordinate systems is a linear transformation,

$$(*) \qquad \begin{aligned} x' &= a_1x + a_2y + a_3, \\ y' &= b_1x + b_2y + b_3, \end{aligned}$$

with $a_1b_2 - a_2b_1 \neq 0$. (We have "absorbed" the multipliers α, β into the coefficients.) Conversely, it is clear that every such linear transformation can be regarded as a change of coordinates. Since a change of coordinates is a change of "names" (no point is "moved"; only its names are changed) it might be called an *alias.* But there is another way of interpreting equations $(*)$. We might regard the coordinate system as being fixed and think of equations $(*)$ as transforming or moving a point P with coordinates (x, y) to a point P' with coordinates (x', y') *in the same coordinate system.* Thus equations $(*)$ effect a motion of the points of Π, and since each point (with perhaps some exceptions) is no longer where it was, equations $(*)$, regarded in this way, might be called an *alibi.* In this light, *an affine property is one that is preserved by every motion.*

The set of linear transformations (*) is easily seen to form a *group*, where the product of two elements of the set is the linear transformation obtained by applying the transformations successively, in a given order. This group is the *affine group of the plane*, and affine plane geometry is simply the study of all those properties of plane figures that are invariant (unchanged) when the figures are operated on (transformed) by every element of this group.

The reader may have noticed that our figures differ from those found in an elementary analytic geometry book in not having arrowheads on the *x*-line and *y*-line. Since the coordinate set Γ of Π (even when the Pappus property is assumed) is not necessarily *ordered*, there is no basis for speaking of positive or negative elements of Γ and hence no ground for assigning positive or negative directions to the coordinate axes. If we desire to "normalize" our geometry to the extent of having the coordinate set the *real number field* (instead of merely a field 𝔉, as insured by Theorem V.8.4) it is necessary and sufficient to impose assumptions that result in 𝔉 being *ordered* and *continuous*.

DEFINITION. *A field* 𝔉 *is an ordered field, provided a binary relation (denoted by* < *and read "precedes") is defined in* 𝔉, *with the following properties:*

O₁ (Trichotomy). *If* a, $b \in \mathfrak{F}$, *exactly one of the relations* $a = b$, $a < b$, $b < a$ *holds.*

O₂ (Transitivity). *If* $a, b, c \in \mathfrak{F}$, *and* $a < b$, $b < c$, *then* $a < c$.

O₃ (Additive monotone). *If* a, b, $c \in \mathfrak{F}$ *and* $a < b$, *then* $a + c < b + c$.

O₄ (Multiplicative semi-monotone). *If* a, b, $c \in \mathfrak{F}$, *then* $a < b$ *and* $0 < c$ *imply* $a \cdot c < b \cdot c$.

If $a \in \mathfrak{F}$, a is called *positive*, provided $0 < a$, and a is called *negative*, provided $a < 0$.

The set of all rational numbers is an ordered field, where < is interpreted as "less than." No finite field can be ordered. The complex number field is an example of an infinite field that cannot be ordered. The real number field is an ordered field, but, as instanced by the rational numbers, there are ordered fields that are not isomorphic to the real number field. The real number field is characterized among all ordered fields by the property of being *continuous*.

DEFINITION.   *An ordered field $\mathfrak{F}$ is continuous, provided whenever $\mathfrak{F}$ is decomposed into two non-empty subsets L and G (that is, $a \in \mathfrak{F}$ implies $a \in L$ or $a \in G$) such that each element of L precedes each element of G, either L contains a greatest element (that is, one that is preceded by every other element of L) or G contains a smallest element (that is, one that precedes every other element of G).*

It is not difficult to show that the real number system is essentially the only continuous ordered field (that is, every such field is isomorphic to it).

Thus if $\mathfrak{F}$ is a continuous ordered field, we have affine plane geometry over the reals. We may obtain ordinary euclidean plane geometry either by introducing postulates of congruence or (equivalently) by considering as motions only the elements of that subgroup of the affine group

$$x' = a_1 x + a_2 y + a_3,$$

$$y' = b_1 x + b_2 y + b_3, \quad a_1 b_2 - a_2 b_1 \neq 0$$

obtained by requiring the coefficients to satisfy the equations $a_1^2 + b_1^2 = a_2^2 + b_2^2 = 1$, $a_1 a_2 + b_1 b_2 = 0$. The three-parameter group thus obtained is usually written in the form $x' = x \cos \theta - y \sin \theta + a$, $y' = x \sin \theta + y \cos \theta + b$ when $a_1 b_2 - a_2 b_1 > 0$.

● EXERCISES

1. Show that no finite field can be ordered.
2. Prove that the field of complex numbers cannot be ordered.

VI

Coordinatizing Projective Planes

Foreword

Having studied the affine plane Π as a postulational system and determined those properties of the plane that permit its coordinate set to form various kinds of algebraic structures (planar ternary rings, Veblen-Wedderburn systems, division rings, and fields), we might proceed in a direct manner to investigate such matters in the projective plane by obtaining that plane from Π by the usual device of adjoining to Π so-called *ideal* points (defined to be *parallel classes of lines*) and one additional line (the *ideal line*) formed by the set of all ideal points. Though this is essentially what our procedure comes to, it seems preferable to start anew, presenting the projective plane as a set of postulates that differ in one important respect from those defining the affine plane. Our study of the affine plane should facilitate greatly our comprehension of the procedures and devices to be employed in this chapter and should permit a more rapid development of the subject.

The duality of point and line, so important in the theory of the projective plane, suggests that each should be regarded as an abstract element. (The reader will recall that in the postulates for an affine plane a line is a class of points.) This necessitates the introduction of *incidence* as a primitive relation.

VI.1. The Postulates for a Projective Plane, and the Principle of Duality

Let Σ denote an abstract set, whose elements are called points and are denoted by capital letters $A, B, \cdots$, and let Λ denote another abstract set, whose elements are called lines and are denoted by lower case letters $a, b, \cdots$. For each element A of Σ and each element a of Λ a relation called incidence is defined, subject to the following postulates:

P_1.   *If P denotes a point and p denotes a line, and P is incident with p, then p is incident with P.*

P_2.   *If p denotes a line and P denotes a point, and p is incident with P, then P is incident with p.*

P_3.   *Two distinct points are together incident with exactly one line.*

P_4.   *Two distinct lines are together incident with exactly one point.*

P_5.   *There are four pairwise distinct points such that no three of them are incident with the same line.*

We shall frequently use "on" as a synonym for "incident," though the reader should avoid conferring any meaning on this term other than that contained in the postulates.

Postulates P_1 through P_5 define a *projective plane*, which may, of course, contain only a finite number of points.

A model of this postulate system is obtained by the following interpretations. Consider the set of all ordered triples of real numbers, *except the triple* $(0, 0, 0)$, and define two triples (x_1, x_2, x_3), (y_1, y_2, y_3) to be equivalent, provided there is a non-zero constant k such that $y_i = k \cdot x_i$ $(i = 1, 2, 3)$. It is clear that this is an equivalence relation, in the technical sense of the term (Section IV.10); hence it decomposes the set of all ordered triples of real numbers [with triple $(0, 0, 0)$ deleted] into *equivalence classes* of ordered triples. Define each equivalence class to be a *point*, which may be represented by any member of the class. Thus (x_1, x_2, x_3) and (kx_1, kx_2, kx_3), $k \neq 0$, $x_1, x_2, x_3 \neq 0, 0, 0$, represent the same point.

Define each of the above equivalence classes to be also a *line* (and each line to be one of these equivalence classes), and define point (x_1, x_2, x_3) to be incident with line (u_1, u_2, u_3) and line (u_1, u_2, u_3) to be incident with point (x_1, x_2, x_3) if and only if

$$u_1x_1 + u_2x_2 + u_3x_3 = 0.$$

With these interpretations, Postulates P_1 and P_2 are obviously true. To establish Postulate P_3 we suppose $A(a_1, a_2, a_3)$ and $B(b_1, b_2, b_3)$ are distinct points, and we consider the equations

$$a_1u_1 + a_2u_2 + a_3u_3 = 0,$$

$$b_1u_1 + b_2u_2 + b_3u_3 = 0.$$

Since not every second-order minor of the matrix

$$\begin{pmatrix} a_1 & a_2 & a_3 \\ b_1 & b_2 & b_3 \end{pmatrix}$$

is zero (otherwise the two ordered number triples belong to the same equivalence class, so $A = B$), the general solution of the equations is

$$u_1 = k(a_2b_3 - a_3b_2), \quad u_2 = k(a_3b_1 - a_1b_3), \quad u_3 = k(a_1b_2 - a_2b_1),$$

where k is an arbitrary constant. Hence there is a unique line incident with the two distinct points A, B.

If $a(a_1, a_2, a_3)$ and $b(b_1, b_2, b_3)$ are distinct lines, the same algebraic argument shows that there is a unique point $x(x_1, x_2, x_3)$ on both lines, so Postulate P_4 is satisfied.

Points $(1, 0, 0)$, $(0, 1, 0)$, $(0, 0, 1)$, $(1, 1, 1)$ fulfill the requirement of postulate P_5.

We shall refer to this model as *the projective plane over the reals*.

A finite model of the postulational system P_1 through P_5 is the finite projective plane 7_3 of Section III.3 (see Figures 5 and 6); another is the finite projective plane of thirteen points and thirteen lines, each line incident with exactly four points and each point incident with exactly four lines, also given in Section III.3.

The unique line incident with two distinct points A, B will be called the *join* of A, B and will be denoted by line AB; the unique point incident with two distinct lines p, q will be referred to as the *meet* or *intersection* of p, q and will be denoted by $p \cdot q$.

If A_1, A_2, A_3, A_4 are four points satisfying the requirement of postulate P_5, they are pairwise distinct, and joining them pairwise we obtain six lines A_iA_j ($i, j = 1, 2, 3, 4; i < j$). Intersecting these lines pairwise yields three additional points B_1, B_2, B_3 with A_1, A_2, B_1 on line p_1; A_1, A_3, B_2 on line p_2; A_1, A_4, B_3 on line p_3; A_2, A_3, B_3 on line p_4; A_2, A_4, B_2 on line p_5; A_3, A_4, B_1 on line p_6 (Figure 33). The seven points are pairwise distinct, as are the six lines. *Such a figure is present in every projective plane.*

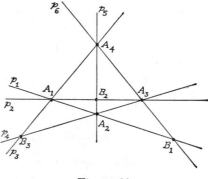

Figure 33

LEMMA VI.1.1.   *Every line is on at least three points.*

Proof.   Each line in Figure 33 is on at least three points. If p is a line distinct from p_i ($i = 1, 2, \cdots , 6$) and A_1 is not on p, by Postulates P_3 and P_4, p meets p_1, p_2, p_3 in three pairwise distinct points. If A_2 is not on p, then p meets p_1, p_4, p_5 in three pairwise distinct points. If neither of these eventualities holds, p is on both A_1 and A_2, and, consequently, p is the same line as p_1. In this case also p is on at least three points.

LEMMA VI.1.2.   *There exist four lines, no three of which are on the same point.*

Proof.   Lines p_1, p_2, p_5, p_6 of Figure 33 have the desired property.

We are now in a position to establish an important property of the postulational system P_1 through P_5 that is lacking in the postulational system defining the affine plane. This property, called the principle of duality, rests on the observation that if "point" and "line" are interchanged in those postulates (thereby obtaining their *duals*), Postulates P_1, P_2, P_3, P_4, and P_5 become Postulates P_2, P_1, P_4, P_3, and Lemma VI.1.2, respectively. Hence the same interchange carried out on any logical consequence of the five postulates (that is, on any theorem) yields a theorem, whose formal proof may be obtained by effecting the interchange in each statement occurring in the proof of the first theorem. Since the validity of each theorem of the system rests ultimately on the five postulates of the system, the validity of the so-called dual of the theorem is based on the duals of the five postulates, which (as we have seen) are Postulates P_1, P_2, P_3, P_4 and Lemma VI.1.2, which is a logical consequence of the original set of postulates.

The validity of the duality principle is of enormous value in the study of projective geometry. The theorems of that subject occur in pairs, allowing for the fact that some theorems are self-dual. In some important instances the dual of a theorem is its converse, which in those cases is automatically valid!

LEMMA VI.1.3. *Every point is on at least three lines.*
Proof. This is the dual of Lemma VI.1.1.

VI.2. Homogeneity of Projective Planes. Incidence Matrices of Finite Projective Planes

In the finite model 7_3 of the postulational system P_1, P_2, $\cdots$, P_5 each line is on exactly three points, and each point is on exactly three lines, and the finite model of thirteen points exhibits the same kind of homogeneous structure, with "three" replaced by "four." We show that every projective plane, finite or infinite, is homogeneous in the sense that if p, q are any two lines, and $I(p)$, $I(q)$ denote the sets of points incident with p, q, respectively, $I(p)$, $I(q)$

have the same cardinal number, which is also the cardinal number of the set of all lines incident with an arbitrary point.

LEMMA VI.2.1. *If P is any point, there is a line not on P.*

Proof. Let p denote any line on P (such a line exists, since the plane has at least seven points), and let Q be a point on p, $P \neq Q$ (Lemma VI.1.1), and let R be a point not on p (Postulate P_5). Then line QR is not on P, for if it is, it coincides with p (Postulate P_3), and R is on p, contrary to its selection.

LEMMA VI.2.2. *If P is any point and q is any line not on P, there is a one-to-one correspondence between the elements of the two sets $i(P)$, $I(q)$, where $i(P)$ denotes the set of all lines incident with P.*

Proof. The desired correspondence is established by associating with each line of $i(P)$ that unique point of $I(q)$ incident with it and q.

LEMMA VI.2.3. *If p, q are any two lines, there is a one-to-one correspondence between the elements of the sets $I(p)$, $I(q)$.*

Proof. Let O denote the unique point incident with both p and q $(p \neq q)$. Select P, Q on p, q, respectively, each distinct from O, and choose a point R on line PQ, distinct from P and Q (Lemma VI.1.1). Then R is not on either p or q, and the one-to-one correspondence between the elements of the sets $I(p)$, $i(R)$, combined with the one-to-one correspondence between the elements of the sets $I(q)$, $i(R)$, insured by Lemma VI.2.2, defines a one-to-one correspondence between the elements of the sets $I(p)$, $I(q)$.

LEMMA VI.2.4. *If P is any point and q any line, there is a one-to-one correspondence between the elements of the sets $i(P)$, $I(q)$.*

Proof. If q is not on P, the result follows from Lemma VI.2.2. If q is on P, let r be a line *not* on P (Lemma VI.2.1). Then the elements of $I(q)$ are in a one-to-one correspondence with those of $I(r)$ (Lemma VI.2.3), and the latter elements are in such a correspondence with the elements of $i(P)$ (Lemma VI.2.2). Hence the sets $I(q)$, $i(P)$ can have their elements put in a one-to-one correspondence.

By the duality principle (applied to Lemma VI.2.3) the sets $i(P)$, $i(Q)$ have the same cardinality and the lemmas combine to yield the following theorem and corollary.

THEOREM VI.2.1. *For each two points P, Q and each two lines s, t, the sets $i(P)$, $i(Q)$, $I(s)$, $I(t)$ have the same cardinality.*

COROLLARY VI.2.1. *If, in a finite projective plane, one line is incident with exactly $n + 1$ points, every line has this property, and every point is incident with exactly $n + 1$ lines.*

A projective plane is said to be of order n, provided one of its lines (and hence *each* of its lines) is incident with exactly $n + 1$ points. Then each point is incident with exactly $n + 1$ lines.

THEOREM VI.2.2. *A projective plane of order n contains exactly $n^2 + n + 1$ points and exactly $n^2 + n + 1$ lines.*

Proof.   Let p be a line and Q a point not incident with it. Each point of the plane is incident with one of the $n + 1$ lines joining Q to each of the $n + 1$ points incident with p. Since each line is incident with exactly $n + 1$ points, the number of points of the plane is clearly $(n + 1)^2 - n = n^2 + n + 1$ [the number n is subtracted from $(n + 1)^2$ because Q is counted $n + 1$ times in the total given by $(n + 1)^2$].

Duality implies that there are exactly $n^2 + n + 1$ lines in a projective plane of order n, since each point of such a plane is incident with exactly $n + 1$ lines.

Remark.   If a projective plane has exactly $n^2 + n + 1$ points, it has exactly $n^2 + n + 1$ lines. If m is the order of the plane, it follows from Theorem VI.2.2 that the plane has exactly $m^2 + m + 1$ points, and, consequently, $m^2 + m + 1 = n^2 + n + 1$. Hence

$(m - n)(m + n + 1) = 0$, and since the second factor does not vanish, $m = n$. So the order of the plane is n, and it has $n^2 + n + 1$ lines.

Consider a projective plane of order n. Label its $n^2 + n + 1$ points $P_1, P_2, \cdots, P_{n^2+n+1}$, its $n^2 + n + 1$ lines $p_1, p_2, \cdots, p_{n^2+n+1}$, in any manner, and construct the table

	P_1	P_2	P_3	$\cdots$	P_N
p_1	a_{11}	a_{12}	a_{13}	$\cdots$	a_{1N}
p_2	a_{21}	a_{22}	a_{23}	$\cdots$	a_{2N}
p_3	a_{31}	a_{32}	a_{33}	$\cdots$	a_{3N}
.	.	.	.	$\cdots$	.
.	.	.	.	$\cdots$	.
.	.	.	.	$\cdots$	.
p_N	a_{N1}	a_{N2}	.	$\cdots$	a_{NN}

where $N = n^2 + n + 1$, $a_{ij} = 1$ or 0, depending on whether line p_i is or is not incident with point P_j, respectively $(i, j = 1, 2, \cdots, N)$. The square matrix $A = (a_{ij})$ $(i, j = 1, 2, \cdots, N)$ is called an *incidence matrix* of the projective plane of order n. The matrix obviously depends not only on the incidence relations exhibited by the points and lines of the plane, but also on the labeling of the elements.

THEOREM VI.2.3. *If A is an incidence matrix of a projective plane of order n, and A^T denotes its transpose (the matrix whose rows and columns are, respectively, the columns and rows of A), $A \cdot A^T = B$, a square matrix of $n^2 + n + 1$ rows, with each element on the principal diagonal equal to $n + 1$, and each element off this diagonal equal to 1.*

Proof. Since each line is on exactly $n + 1$ points, each row of A has 1 in exactly $n + 1$ places. Dually, since each point is incident with exactly $n + 1$ lines, each column of A has 1 in exactly $n + 1$ places. Writing $B = (b_{ij})$ $(i, j = 1, 2, \cdots, n^2 + n + 1)$,

$$b_{ii} = \sum_{k=1}^{N} a_{ik} a_{ik} = \sum_{k=1}^{N} a^2_{ik} = n + 1 \; (i = 1, 2, \cdots, N = n^2 + n + 1).$$

Now, for each two distinct indices i, j, clearly $a_{ik}a_{kj} = 1$ for exactly one value of k, and, consequently

$$b_{ij} = \sum_{k=1}^{N} a_{ik}a_{jk} = 1, \ (i, j = 1, 2, \cdots, N; \ i \neq j),$$

which proves the theorem.

COROLLARY VI.2.3. *Every incidence matrix A is non-singular; that is, $|A| \neq 0$, where $|A|$ denotes the determinant of A.*

Proof. Clearly, $|A|^2 = |AA^T| = |B| = (n + 1)^2 n^{n(n+1)}$; hence $|A| = \pm(n + 1)n^{n(n+1)/2} \neq 0$, since $n \geq 2$.

Remark. If A is an incidence matrix of a finite projective plane, $AA^T = A^TA$. This follows from the fact that each column of A has 1 in exactly $n + 1$ places and that in any two columns of A exactly one row consists entirely of 1's.

THEOREM VI.2.4. *If a matrix A of non-negative integers and order $n^2 + n + 1(n \geq 2)$ satisfies the equations*

$$(*) \qquad\qquad AA^T = A^TA = B$$

(where B is the matrix defined in Theorem VI.2.3), each element of A is either 0 or 1, and A is an incidence matrix of a finite projective plane of order n.

Proof. Assume that an element $a_{\alpha\beta}$ of A exceeds 1. The element in row i and column α of AA^T is given by

$$a_{i1}a_{\alpha1} + a_{i2}a_{\alpha2} + \cdots + a_{i\beta}a_{\alpha\beta} + \cdots + a_{iN}a_{\alpha N} = 1 \text{ for } i \neq \alpha.$$

Since $a_{\alpha\beta} > 1$, $a_{i\beta} = 0$ for every index i different from α.

The element in row β and column i of A^TA is given by

$$a_{1\beta}a_{1i} + a_{2\beta}a_{2i} + \cdots + a_{\alpha\beta}a_{\alpha i} + \cdots + a_{N\beta}a_{Ni} = 1$$

for $i \neq \beta$. Hence $a_{\alpha i} = 0$ for $i \neq \beta$.

Now the element in row α and column i of AA^T is given by

$$a_{\alpha1}a_{i1} + a_{\alpha2}a_{i2} + \cdots + a_{\alpha\beta}a_{i\beta} + \cdots + a_{\alpha N}a_{iN} = 0$$

for $i \neq \alpha$, so matrix A does not satisfy equations $(*)$.

Thus, each element of A is either 0 or 1, and it follows quite easily that each row and each column of A has 1 in exactly $n + 1$ places and that in any two rows (columns) of A exactly one column (row) consists exclusively of 1's. Consequently, A is an incidence matrix of a projective plane of order n.

The problem of determining all of those integers n for which projective planes of order n exist is unsolved. Use of Theorems VI.2.3 and VI.2.4 make it possible to employ large computing machines to examine conjectured values of n. For every $n = p^k$ (p, a prime, k, a natural number), projective planes of order n exist. On the other hand, it has been shown that projective planes of order n do *not* exist for $n = 2p$, where p is a prime of the form $4m + 3$ (m, a natural number). The smallest natural number n for which the problem is undecided is $n = 10$.

● EXERCISES

1. Construct an incidence matrix A for the projective geometry 7_3.
2. Find the matrix $B = AA^T$ and evaluate it, where A is the matrix in Exercise 1.

VI.3. Introduction of Coordinates

Let A, B, O, I be four points, no three of which are incident with the same line (postulate P_5), and consider that one-to-one correspondence γ' of the elements of the set $i(A)$ of lines on A, with the elements of the set $i(B)$ of lines on B, obtained by associating a line of $i(A)$ with a line of $i(B)$, provided the two lines intersect on line OI (Figure 34).

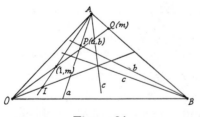

Figure 34

Let Γ denote any abstract set whose elements are in a one-to-one correspondence γ with the elements of the set $i'(A)$ (read "i apostrophe of A"), where $i'(A)$

denotes the set obtained by deleting from $i(A)$ the line AB, and let two distinct elements of Γ be labeled 0, 1, respectively. A one-to-one correspondence γ^* of the elements of Γ with those of $i'(B)$, the set obtained by deleting line AB from $i(B)$, is established by associating with each line of $i'(B)$ that element of Γ that is associated with that line of $i'(A)$ corresponding to it by γ'. From now on, lower-case letters $a, b, \cdots, x, y$ denote elements of the *coordinate set* Γ, and a line will be denoted by writing in juxtaposition the symbols of two of its points; for example, PQ. We let $\gamma(AO) = 0 = \gamma^*(BO)$ and $\gamma(AI) = \gamma^*(BI) = 1$.

Suppose P is any point *not* on AB. The ordered pair (a, b) of elements of Γ is attached to P as *coordinates*, where $\gamma(AP) = a$ and $\gamma^*(BP) = b$. Conversely, each ordered pair (c, d) of elements of Γ is the coordinate pair of a unique point not on AB—the point in which lines AP, BQ of $i'(A)$, $i'(B)$, respectively, intersect, where $\gamma(AP) = c$ and $\gamma^*(BQ) = d$. Hence there is a one-to-one correspondence between the set of all points of the plane *not incident with AB* and the set of all ordered pairs of elements of Γ. Clearly, each point on AO, distinct from A, has its first coordinate 0; each point on BO, distinct from B, has its second coordinate 0; and each point on OI has its first coordinate equal to its second.

To assign coordinates to points Q on AB, $Q \neq A$, we observe that OQ intersects AI in a point with coordinates $(1, m)$, $m \in \Gamma$. We take the second coordinate m of this point as the sole coordinate of Q and say that Q has coordinate (m). Conversely, for each element d of Γ there is a unique point D of AB, $D \neq A$, with coordinate (d). It is the intersection of AB and the join of the points O, $P(1, d)$ [since $P(1, d)$ is not incident with AO, point $D \neq$ point A].

Thus each point of the plane has been assigned coordinates from the set Γ, *except the point A*, to which *no* coordinates are assigned.

● EXERCISES

1. Verify that coordinates of O and I are $(0, 0)$ and $(1, 1)$, respectively, and that B has coordinate (0).

2. Coordinatize the projective plane 7_3.

VI.4. The Ternary Operation in Γ.
Addition and Multiplication

A ternary operation $T(x, m, b)$ is defined in Γ by associating with the ordered triple $x, m, b \in$ Γ the *second coordinate* of the point in which the line of $i'(A)$ that corresponds by γ to x intersects the join of $Q(m)$ and the point $(0, b)$ of AO (Figure 35). [$Q(m)$ denotes the unique point of AB with coordinate (m).]

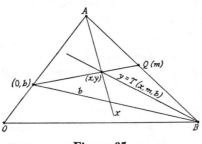

Figure 35

It is clear that the ternary operation depends on the choice of the *four base points* A, B, I, O; for different choices of these points, different (in general) ternary operations are defined in Γ. *Every set of four points, no three of which are incident with the same line, determines a ternary operation T with properties (1) through (7) of Section IV.6.* (Let the reader show this.) Two additional properties are:

(8) If $m \neq 0$, the equation $T(x, m, b) = c$ has a unique solution.

(9) If $a \neq 0$, the equation $T(a, x, b) = c$ has a unique solution, for all $b, c \in$ Γ.

By property *(5)* of Section IV.6, if $m \neq 0$, the equation $T(x, m, b) = T(x, 0, c)$ has exactly one solution. Since, by property *(3)*, $T(x, 0, c) = c$, property *(8)* follows. To prove property *(9)*, consider the system of equations $T(a, x, y) = c$, $T(0, x, y) = b$, $a \neq 0$. By property *(6)*, this system has a unique solution. The second equation gives $y = b$, by property *(3)*, so $T(a, x, b) = c$ has a unique solution.

The system $[\Gamma, T]$ is called a planar ternary ring, and *any* set S in which a ternary operation t is defined having properties *(1)* through *(7)* [hence properties *(8)* and *(9)* also] is called a *ternary system*.

THEOREM VI.4.1. *If* $[S, t]$ *is any ternary system, it deter-mines a projective plane whose points are (i) ordered pairs* (a, b) *of elements of* S, *(ii) elements* (m), $m \in S$ *and (iii) a symbol* A, *and whose lines are: equations* $y = t(x, m, b)$; *equations* $x = a$; *and a symbol* $\mathfrak{L}_\infty$; *where* a, b, m *are elements of* S. *The incidence relation is defined as follows:*

   (i) (a, b) *is incident with* $y = t(x, m, c)$ *if and only if* $b = t(a, m, c)$;

   (ii) (a, b) *is incident with* $x = a$ *for every* b;

   (iii) *point* (m) *is incident with* $y = t(x, m, b)$ *for every* b;

   (iv) *point* (m) *is incident with* $\mathfrak{L}_\infty$ *for every* m;

   (v) *point* A *is incident with* $\mathfrak{L}_\infty$ *and with* $x = a$ *for every element* a.

The reader is asked to prove this theorem in Exercise 1 below.

DEFINITION. *If* $a, b \in \Gamma$, $a + b = T(a, 1, b)$ *and* $ab = T(a, b, 0)$.

● EXERCISES

1. Prove Theorem VI.4.1. Show that in the projective plane determined by the ternary system $[S, t]$, points A, $B = (0)$, $O = (0, 0)$, and $I = (1, 1)$ are base points of a coordinate system with respect to which $[S, t]$ is the planar ternary system.

2. Prove geometrically that in any planar ternary system $[\Gamma, T]$

   (1) $a \cdot 1 = 1 \cdot a = a$, all a of Γ,

   (2) $a + 0 = 0 + a = a$, all a of Γ,

   (3) $a \cdot 0 = 0 \cdot a = 0$, all a of Γ.

3. If $a, b \in \Gamma$, give geometrical constructions for $a + b$ and $a \cdot b$.

4. Show that the equation $a + x = c$ has a unique solution $(a, c \in \Gamma)$.

5. Show that if $a \neq 0$ and $b \neq 0$, each of the equations $xb = c$, $ay = c$ has a unique solution $(a, b, c \in \Gamma)$.

VI.5. Configurations

A set of a_{11} points and a_{22} lines, together with an incidence relation, forms a *configuration*, provided each point is on the same number a_{12} of lines and each line is on the same number a_{21} of points. Such a figure is conveniently described by a square matrix

$$\begin{pmatrix} a_{11} & a_{12} \\ a_{21} & a_{22} \end{pmatrix}.$$

Each finite projective plane is a configuration, with $a_{11} = a_{22} = n^2 + n + 1$, where n is the order of the plane and $a_{12} = a_{21} = n + 1$. In particular, the finite projective plane 7_3 has the symbol

$$\begin{pmatrix} 7 & 3 \\ 3 & 7 \end{pmatrix}.$$

It is an important problem to determine those projective planes in which given configurations are realized. An instance is given in the following theorem.

THEOREM VI.5.1.   *The configuration with symbol*

$$\begin{pmatrix} 7 & 3 \\ 3 & 7 \end{pmatrix}$$

is realized in a projective plane if and only if in each of its planar ternary rings $[\Gamma, T]$, $a + a = 0$ *for every element a of Γ.*

Proof.   Suppose $a + a = 0$ for every element a of every planar ternary ring of the plane, and let the marks *1, 2, 3,* $\cdots$ *, 6, 7* denote seven points of the plane such that

$$\left.\begin{matrix} 1 & 3 \\ 2 & 4 \end{matrix}\right\} 5 \quad \left.\begin{matrix} 1 & 2 \\ 3 & 4 \end{matrix}\right\} 6 \quad \left.\begin{matrix} 1 & 4 \\ 2 & 3 \end{matrix}\right\} 7,$$

where the notation indicates that point *5* is the intersection of lines *13* and *24*, and so on. We wish to show that points *5, 6, 7* are incident with the same line.

Choose a coordinate system with $A = 1$, $B = 2$, $O = 3$, and I on line *36* ($I \neq 3, 6$) (Figure 36).

If $\gamma(17) = a$, point *4* has coordinates (a, a) and point *5* has coordinates $(0, a)$. Since the coordinate of *6* is (1), line *56* has equation $y = T(x, 1, a) = x + a$, and since $a + a = 0$, the coordinates $(a, 0)$ of point *7* satisfy this equation, and, consequently, point *7* is on line *56*.

Conversely. if points *5, 6, 7* are on the same line, then $y = x + a$

(an equation of line *56*) must be satisfied by $(a, 0)$, the coordinates of point *7*, so $a + a = 0$.

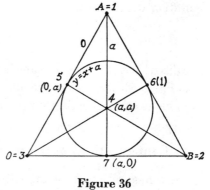

Figure 36

Remark. Let P_1, P_2, P_3, P_4 be four points of the projective plane over the reals (Section VI.1), no three of the points being incident with the same line. These four points, together with the *six* pairwise distinct lines obtained by joining them in pairs, form a *complete quadrangle*. Two lines of the complete quadrangle are called *opposite* if their intersection is *not* one of the four points P_1, P_2, P_3, P_4. The intersection of two opposite lines is called a *diagonal point* of the complete quadrangle.

In Figure 36 points *5*, *6*, *7* are the three diagonal points of the complete quadrangle determined by the four points *1*, *2*, *3*, *4*.

The homogeneous coordinatization of the projective plane over the reals can be made non-homogeneous in the following way. All those points (x_1, x_2, x_3) with $x_3 \neq 0$ have non-homogeneous coordinates $(x_1/x_3, x_2/x_3)$, whereas points $(x_1, x_2, 0)$ have the single coordinate x_2/x_1, if $x_1 \neq 0$. No coordinate is assigned to the point with homogeneous coordinates $(0, x_2, 0)$, $x_2 \neq 0$. The reader will recognize that the projective plane over the reals is now coordinatized in a manner entirely analogous to the way in which the abstract projective plane was coordinatized in Section VI.3. Here, Γ is the set of real numbers and the ternary operator $T(x, m, b) = m \cdot x + b$, where the $\cdot$ and $+$ indicate ordinary multiplication and addition, respectively. Since it is false that $a + a = 0$ for every real number a, we have the following result as a corollary of Theorem VI.5.1.

COROLLARY VI.5.1. *In the projective plane over the reals the diagonal points of a complete quadrangle are never collinear.*

A projective plane in which the diagonal points of each complete quadrangle are collinear [that is, a projective plane in which a configuration with symbol

$$\begin{pmatrix} 7 & 3 \\ 3 & 7 \end{pmatrix}$$

is realized] is called a Fano plane, after the Italian mathematician, Gino Fano, who worked in the foundations of geometry during the late nineteenth and early twentieth centuries.

VI.6. Configurations of Desargues and Pappus

The configuration of Desargues has the symbol

$$\begin{pmatrix} 10 & 3 \\ 3 & 10 \end{pmatrix};$$

that is, it consists of ten points and ten lines, each line on three points and each point on three lines. A projective plane has the Desargues property if the Desargues configuration is realizable in that plane (that is, if the Desargues theorem is valid in the plane). The theorem is usually stated as follows:

THEOREM OF DESARGUES. *If the three pairs of corresponding vertices of two triangles are joined by three lines on the same point, the three pairs of corresponding sides of the triangle intersect in three points on the same line.*

Remark 1. The dual of Desargues' theorem is its converse, hence the theorem is valid in a projective plane if and only if its converse is valid.

Remark 2. The reader should observe that the specializations of the above theorem, that gave rise to the first and third Desargues properties in affine geometry (see the Foreword to Chapter V and Section V.6), are *not à propos* in the present context. Those specializations referred to parallelism of certain lines, a relation that never occurs in the projective plane (postulate P_4). It is occasionally useful to regard two lines that intersect on line AB as being "parallel," but this is a property that is relative to a particular coordinate system (not an intrinsic property of the two lines) and hence is without geometric significance.

Remark 3. Projectively significant special cases of Desargues' theorem arise on restricting one, two, or even three vertices of one triangle to be on the sides of the other. Let D_1 denote the Desargues theorem when exactly one vertex of one triangle is assumed to be on exactly one side of the other triangle.

Calling two triangles *perspective with respect to a point (line)*, provided the joins (intersections) of the three pairs of corresponding vertices (sides) are on the point (line), Desargues' theorem states that if two triangles are perspective with respect to a point, they are perspective with respect to a line. A special case, known as the *minor theorem* (or property) of Desargues, asserts that if two triangles are perspective with respect to a point P, and if two pairs of corresponding sides intersect in points of a line p *on* P, the third pair of corresponding sides intersects on that line. We shall show later that this minor theorem is equivalent to D_1. The first Desargues property is an affine specialization of the minor Desargues property.

Not every projective plane has the Desargues property. The affine plane defined in Section IV.11 can be made into a projective plane in the manner described in the Foreword of this chapter, adjoining to each broken euclidean line the *direction of the upper half* as the ideal point on that line. Figure 37 shows that the converse of the Desargues theorem is not valid in that plane, and, consequently, neither is the theorem itself (Remark 1).

The triangles A_1, B_1, C_1 and A_2, B_2, C_2 are perspective with respect to the ideal line of the plane, but they are not perspective with respect to any point. Lines B_1B_2 and C_1C_2 meet at P, but line A_1A_2 (a euclidean broken line) is not on P.

The Pappus configuration has the symbol

$$\begin{pmatrix} 9 & 3 \\ 3 & 9 \end{pmatrix};$$

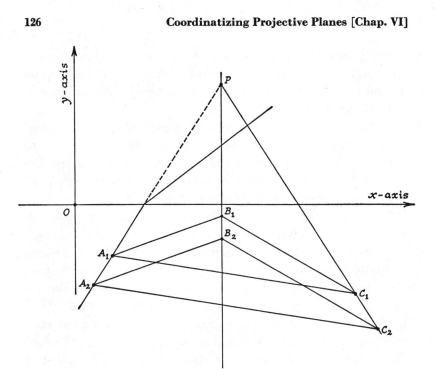

Figure 37

that is, it consists of nine points and nine lines, each point on three lines and each line on three points (Figure 38).

Points *1*, *3*, *5* on one line and points *2*, *4*, *6* on another line are joined "crosswise" to yield points *7*, *8*, *9* (that is, lines *12* and *45* meet at point *7*, lines *23* and *56* meet at point *8*, and lines *34* and *16* meet at point *9*). A plane has the Pappus property, provided points *7*, *8*, *9* are on the same line. The Pappus properties of Section V.7 are affine specializations.

The projective plane that we have just shown to be non-Desarguesian (that is, does not have the Desargues property) does not have the Pappus property either. Let the reader construct a figure and prove this.

Remark. An important and beautiful theorem of classical projective geometry (over the reals) was proved in 1640 by the French mathematician-philosopher, Blaise Pascal (1623–1662). It asserts

that if any six pairwise distinct points of any conic are designated
1, 2, 3, 4, 5, 6, the three pairs *12* and *45*, *23* and *56*, and *34* and *61*
of "opposite" sides of the simple hexagon obtained by joining the
points in order intersect in three points of a line. The theorem of
Pappus results on applying Pascal's theorem to a degenerate conic
consisting of two distinct lines, labeling three points of one line
1, 3, 5 and three points of the other *2, 4, 6*.

Consider the special case of the Pappus property given in Section
V.7, and let P' be the *ideal* point on the line $Q'R'$, that is, P' is the
intersection of line $Q'R'$ with *an arbitrarily chosen fixed line* of the
plane. If *1, 2, 3, 4, 5, 6* are new labels for the points R', Q, P', R, Q',
P, respectively, where P is the ideal point on line QR, the Pappus
theorem applied to this sextuple, and *interpreted affinely*, is called
the *affine minor theorem of Pappus*. It may be stated as follows:
If *1, 5* are distinct points of one line, *2, 4* distinct points of another
line, with line *(1, 2)* parallel to line *(4, 5)*, and if *8* denotes the intersection of the line on *1*, parallel to line *(2, 4)*, with the line on *4*,
parallel to line *(1, 5)*, while *9* is the intersection of the line on *5*,

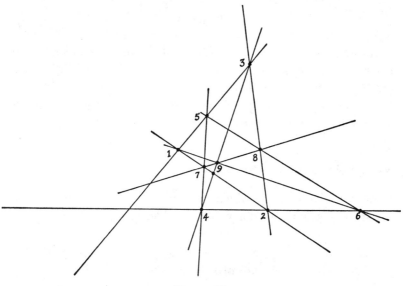

Figure 38

parallel to line (*2, 4*), with the line on *2*, parallel to line (*1, 5*), then
line (*8, 9*) is parallel to line (*1, 2*). [If the ideal point on line (*1, 2*)
and line (*4, 5*) be labeled *7*, the theorem asserts that *7, 8, 9* are on
the same line.] See Figure 39, which is used also in the proof of
Lemma VI.7.1.

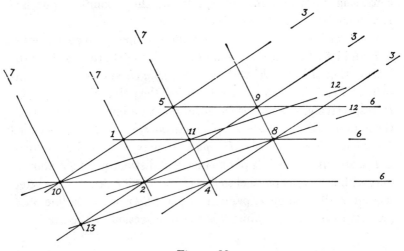

Figure 39

Note that two lines are mutually parallel, provided they intersect
on the arbitrarily selected fixed line. Denote this fixed line by $\mathcal{L}_\infty$.
[In Figure 39, $\mathcal{L}_\infty$ is selected as the infinitely distant line of the extended euclidean plane.] Affine satisfaction of a theorem means
satisfaction with respect to $\mathcal{L}_\infty$ as the locus of ideal points.

VI.7. Veblen-Wedderburn Planes

A projective plane in which the *minor* theorem of Desargues is
satisfied *affinely* is called a Veblen-Wedderburn plane.

LEMMA VI.7.1.   *In a Veblen-Wedderburn plane the minor the-
orem of Pappus is satisfied affinely, with points 3, 6, 7 on $\mathcal{L}_\infty$.*
    Proof.   Put

$$\left.\begin{matrix} 2 & 4 \\ 1 & 3 \end{matrix}\right\} 10 \qquad \left.\begin{matrix} 1 & 6 \\ 4 & 5 \end{matrix}\right\} 11 \qquad \left.\begin{matrix} 3 & 6 \\ 10 & 11 \end{matrix}\right\} 12 \qquad \left.\begin{matrix} 7 & 10 \\ 2 & 3 \end{matrix}\right\} 13.$$

(See Section VI.5 for notation. Note that *10, 11, 12, 13* denote *points*. Refer to Figure 39, where the ideal point *P* on a line *p* is indicated by <u> p </u> <u> P </u> . Since the figure is drawn with $\mathcal{L}_\infty$ chosen as the infinitely distant line in the extended euclidean plane, parallel lines are parallel in the ordinary sense.)

Applying the affine specialization of the minor theorem of Desargues (first Desargues property, Chapter V Foreword) to the triangles *2, 4, 13* and *1, 11, 10*, we conclude that points *4, 12, 13* are on the same line. Now triangles *10, 2, 13* and *11, 8, 4* are such that line (*10, 2*) is parallel to line (*11, 8*), line (*2, 13*) is parallel to line (*8, 4*), line (*10, 13*) is parallel to line (*11, 4*), and lines (*10, 11*) and (*4, 13*) are on point *12*.

By the converse of the minor theorem of Desargues (which follows from it), line (*2, 8*) is also on point *12*.

Finally, apply the minor theorem of Desargues to triangles *5, 10, 11* and *9, 2, 8*. Line (*5, 9*), line (*2, 10*), line (*8, 11*) are pairwise mutually parallel, line (*5, 10*) is parallel to line (*2, 9*), and line (*10, 11*) is parallel to line (*2, 8*). Hence line (*5, 11*) is parallel to line (*8, 9*), and since, by hypothesis, line (*4, 5*) [which is identical with line (*5, 11*)] is parallel to line (*1, 2*), line (*8, 9*) is parallel to line (*1, 2*), and the theorem is proved.

LEMMA VI.7.2. *If the minor theorem of Pappus is satisfied affinely in a projective plane* Π, *and* [Γ, *T*] *is a planar ternary ring of* Π *based on a coordinate system O, I, A, B, with points A, B on* $\mathcal{L}_\infty$, *then* [Γ, *T*] *is an abelian (commutative) group with respect to addition.*

Proof. Let *a, b* be any two distinct elements of Γ, and choose *1, 2, 3, 4, 5* as labels of the points (0, *b*), (*b, b*), *A*, (*a, a*), (0, *a*), respectively (Figure 40).

Line (*1, 2*) has equation $y = b$, and line (*4, 5*) has equation $y = a$. Hence their intersection is point *B*, which we label also *7*. Then line (*3, 7*) is the special line $\mathcal{L}_\infty$. Its intersection with line

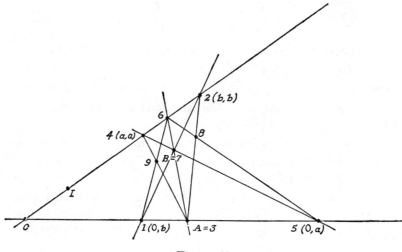

Figure 40

($2, 4$) (whose equation is $y = x$) we label 6. It has coordinate (1).

Line ($2, 3$) has equation $x = b$, and line ($5, 6$) has equation $y = T(x, 1, a) = x + a$. Hence point 8, the intersection of these two lines, has coordinates ($b, b + a$).

Line ($3, 4$) has equation $x = a$, and line ($6, 1$) has equation $y = T(x, 1, b) = x + b$. Hence their intersection, point 9, has coordinates ($a, a + b$).

Now since the minor theorem of Pappus is affinely satisfied, points $7, 8, 9$ are on the same line; that is, points $8, 9$ are on a line with B. Consequently, the second coordinates of points $8, 9$ are equal, so $a + b = b + a$. Hence addition in (Γ, T) is commutative.

This result, combined with Exercise 4 of Section VI.4, shows that equations $a + x = b, x + a = b$ have unique solutions, and it ˜emains only to establish the associativity of addition.

Let $1, 2, 3, 4, 5$ be the labels of points ($0, a + b$), ($a, a + b$), A, ′$c, c + b$), ($0, c + b$), respectively (Figure 41).

Equations of lines ($1, 2$) and ($4, 5$) are $y = a + b$ and $y = c + b$, respectively, so their intersection, point 7, is point B of the coordinate system. Then line ($3, 7$) is $\mathcal{L}_\infty$. By the geometrical construction of addition (Exercise 3, Section VI.4), both points 2 $(a, a + b)$ and 4 $(c, c + b)$ are on the line joining $(0, b)$ to (1); thus, the intersection of lines $(3, 7)$ and $(2, 4)$ is (1). Then line $(5, 6)$ has equation $y = T(x, 1, c + b) = x + (c + b)$, and line $(6, 1)$ has equation $y = T(x, 1, a + b) = x + (a + b)$. Intersecting the first line with line $(2, 3)$ (equation $x = a$) gives 8 $(a, a + (c + b))$; intersecting the second line with line $(3, 4)$ (equation $x = c$) gives 9 $(c, c + (a + b))$.

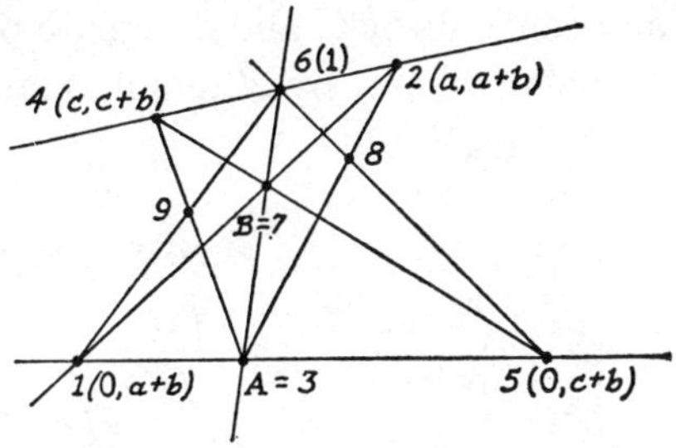

Figure 41

Since points 8, 9 are on a line which is on $7 = B$, we have $a + (c + b) = c + (a + b) = (a + b) + c$, the second equality following from the commutativity of addition just established. Since $a + (c + b) = a + (b + c)$, we obtain $a + (b + c) = (a + b) + c$, and associativity of addition is proved.

THEOREM VI.7.1. *Every planar ternary ring $[\Gamma, T]$ of a Veblen-Wedderburn projective plane, based on a coordinate system O, I, A, B, where A, B are on the special line $\mathcal{L}_\infty$, has the following properties:*

(1) $T(a, m, b) = a \cdot m + b$;

(2) $a + b = b + a$;

(3) $a + (b + c) = (a + b) + c$;

(4) $a + x = c$ *has a unique solution;*

(5) $a \cdot 0 = 0 = 0 \cdot a$;

(6) $a \cdot 1 = a = 1 \cdot a$;

(7) if $a \neq 0$, $a \cdot y = c$ *has a unique solution;*

(8) if $a \neq 0$, $x \cdot a = c$ *has a unique solution;*

(9) if $r \neq s$, $xr = xs + b$ *has a unique solution;*

(10) $(a + b) c = ac + bc$.

Proof. By Lemma VI.7.1, the minor theorem of Pappus is affinely satisfied in the plane with points *3*, *6*, *7* on $\mathcal{L}_\infty$. Hence Lemma VI.7.2 gives properties (2), (3), and (4).

To prove property (1), the linearity of operator T, let a, m, b be any elements of Γ, and designate point $(0, b)$ by *1*, the origin O by *1'*, point (a, am) by *2'*, point A by *4*, point (1) by *6*, and point B by *7* (Figure 42).

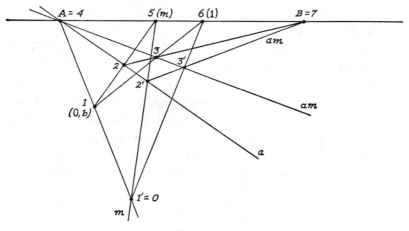

Figure 42

(Let the reader show that the figure may be generated by the points *1*, *1'*, *2'*, *4*, *6*, *7*, and hence that the elements a, m, b are any whatever.)

It follows from the geometrical construction of a product (Exercise 3, Section VI.4) that line (*1'*, *2'*) intersects line (*4*, *7*) in *5*(*m*), and it is clear that *3'* has coordinates (am, am) and that line (*1*, *6*) has equation $y = x + b$. Hence point *3* has coordinates $(am, am + b)$. Line (*1*, *5*) has equation $y = T(x, m, b)$, so point *2* has coordinates $(a, T(a, m, b))$.

Now, since the minor Desargues theorem is affinely satisfied, points *2*, *3*, *7* are on the same line, and since *7* = B, the second coordinates of points *2* and *3* are equal; that is $T(a, m, b) = am + b$.

Properties (5), (6), (7), and (8) are the same as those stated in

Exercises 2 and 5 of Section VI.4, and property (9) follows from property *(5)* of Section IV.6 (which the reader was asked to establish for the ternary operator T defined in Section VI.4) and the linearity of T just established. It remains to prove the distributive property, property (10).

Let a, b, c be any elements of Γ. Designate points B, (c), (1), A, $(b, 0)$, O by *4, 5, 6, 7, 1, 1'*, respectively, and consider line a on A (Figure 43).

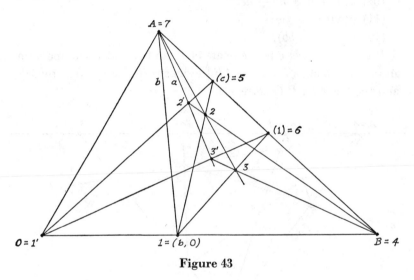

Figure 43

That line meets line *(1', 6)* in point *3'*, with coordinates (a, a), and line *(1', 5)* in point *2'*, with coordinates (a, ac). Line *(3', 4)* meets line *(1, 6)* in point *3*, with coordinates $(a + b, a)$, and line *(2', 4)* meets line *(3, 7)* in point *2*, with coordinates $(a + b, ac)$.

By property (1), line *(1, 5)* has equation $y = xc + d$, for some element d of Γ, and since $(b, 0)$ must satisfy this equation, $0 = bc + d$, so $d = -(bc)$, where $-(bc)$ is the unique additive inverse of the element bc. Hence line *(1, 5)* has equation $y = xc + [-(bc)] = xc - (bc)$. Since point *2* is on line *(1, 5)* (by the minor Desargues theorem applied to triples *1, 2, 3* and *1', 2', 3'*), $ac = (a + b)c - (bc)$. Consequently, $(a + b)c = ac + bc$, and the theorem is proved.

VI.8. Alternative Planes

A projective plane in which the minor theorem of Desargues is satisfied (not *merely affinely satisfied*) is called an *alternative plane.*

THEOREM VI.8.1.   *Besides properties (2)–(8) and (10) of Theorem VI.7.1, addition and multiplication in a planar ternary ring of an alternative projective plane satisfy*

(9*) $a(b + c) = ab + ac$;

(11) $a(ab) = (aa)b$;

(12) $(ab)b = a(bb)$.

Proof.   Let a, b, c be elements of Γ, and consider the line a on $A = 4$, the point (b) on line AB [designate (b) by 1 and B by 1′], and the line c on B (Figure 44).

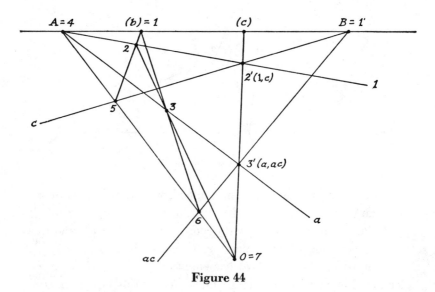

Figure 44

Let O, $(1, c)$, and $(a, ab + ac)$ have labels 7, 2′, and 3, respectively.   The reader will readily verify that point 3′ has coordinates (a, ac) and that the coordinates of point 3 [the intersection of line $(1, 6)$ and line a] are $(a, ab + ac)$.

Point *5* has coordinates $(0, c)$, so line $(1, 5)$ has equation $y =$

$xb + c$. This line intersects line $(2', 4)$ in point 2, which, consequently, has coordinates $(1, b + c)$. It follows that line $(2, 7)$ has
equation $y = x(b + c)$. Since the minor theorem of Desargues is
satisfied in the plane, line $(2, 7)$ is on point 3 (why?), so coordinates
of point 3 are $(a, a(b + c))$. Hence $a(b + c) = ab + ac$.

ASSERTION 1.   *Let the minor theorem of Desargues be satisfied in
a projective plane* II. *If two triangles of* II *are perspective with respect
to a point, and exactly one vertex of one of the triangles is on exactly
one side of the other, the two triangles are perspective with respect to a
line. Conversely, this implies the minor theorem of Desargues.*

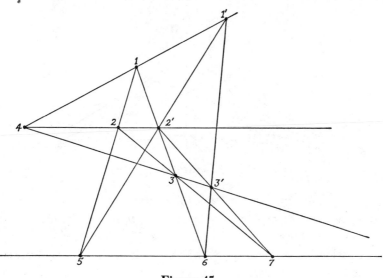

Figure 45

In Figure 45 triangles $(1, 2, 3)$ and $(1', 2', 3')$ are perspective
with respect to point 4, and point $2'$ is on side $(1, 3)$. Lines $(2, 3)$
and $(2', 3')$ meet at point 7, lines $(1, 3)$ and $(1', 3')$ meet at point 6,
and lines $(1', 2')$ and $(6, 7)$ meet at point 5. Triangles $(1', 3', 4)$
and $(5, 7, 2)$ are perspective with respect to point $2'$. Lines $(1', 3')$
and $(5, 7)$ meet at point 6, lines $(3', 4)$ and $(7, 2)$ are on point 3, and
line $(3, 6)$ is on point $2'$. From the minor theorem of Desargues we

conclude that line $(1', 4)$ meets line $(2, 5)$ on line $(3, 6)$, and hence at point 1, which means that triangles $(1, 2, 3)$ and $(1', 2', 3')$ are perspective with respect to the line on points $5, 6, 7$.

Let the reader supply the proof of the converse.

ASSERTION 2. *Let* $1, 2, 3, 4, 5$ *be five points of a projective plane in which the minor theorem of Desargues is satisfied. If*

$$\left.\begin{matrix} 1 & 4 \\ 3 & 5 \end{matrix}\right\} 7 \quad \left.\begin{matrix} 3 & 4 \\ 1 & 5 \end{matrix}\right\} 8 \quad \left.\begin{matrix} 2 & 8 \\ 1 & 4 \end{matrix}\right\} 9 \quad \left.\begin{matrix} 2 & 4 \\ 3 & 9 \end{matrix}\right\} 6 \quad \left.\begin{matrix} 2 & 5 \\ 1 & 4 \end{matrix}\right\} 10 \quad \left.\begin{matrix} 3 & 10 \\ 1 & 6 \end{matrix}\right\} 11,$$

points $2, 7, 11$ *are on the same line* (Figure 46).

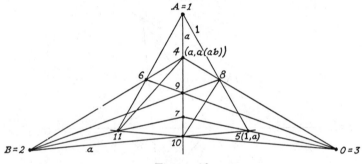

Figure 46

Apply Assertion 1 to triangles $(8, 9, 10)$ and $(4, 6, 11)$ [observing that vertex 4 of one triangle is on side $(9, 10)$ of the other]. It follows that lines $(8, 10)$ and $(4, 11)$ meet on line $(1, 2)$.

Considering triangles $(5, 8, 10)$ and $(7, 4, 11)$ [noting that vertex 10 is on side $(4, 7)$], we have lines $(5, 8)$ and $(4, 7)$ meeting at point 1, and lines $(8, 10)$ and $(4, 11)$ meeting on line $(1, 2)$; consequently, line $(5, 10)$ meets line $(7, 11)$ on line $(1, 2)$. Since line $(5, 10)$ meets line $(1, 2)$ at point 2, line $(7, 11)$ is on point 2, and the assertion is proved.

To prove that $a(ab) = (aa)b$ we may suppose $a \neq 0$, $b \neq 0$. Let $1 = A$, $2 = B$, $3 = O$, $4 = (a, a(ab))$, and $5 = (1, a)$ (Figure 46). Point 7 is the intersection of line $(1, 4)$ (with equation $x = a$) and line $(3, 5)$ (with equation $y = xa$) and hence has coordinates (a, aa). Point 8 is the intersection of line $(3, 4)$ [with equation

$y = x(ab)$] and line $(1, 5)$ (with equation $x = 1$) and hence has
coordinates $(1, ab)$. Point 9, the intersection of lines $(2, 8)$ and
$(1, 4)$, has coordinates (a, ab); point 6, the intersection of lines
$(2, 4)$ and $(3, 9)$ [with respective equations $y = a(ab)$ and $y = xb$],
has coordinates $(c, a(ab))$, where $c \in \Gamma$ such that $(*)$ $cb = a(ab)$
(since $b \neq 0$, c is unique, by property (8), Theorem VI.7.1).
Clearly, point 10 has coordinates (a, a), hence line $(3, 10)$ has equation $y = x$. Since line $(1, 6)$ has equation $x = c$, point 11 has coordinates (c, c).

By Assertion 2, points 2, 7, 11 are on the same line, and, since
this line is on B, its equation is $y = $ const. Hence the second coordinates of points 7 and 11 are equal; that is, $aa = c$. Substituting
in $(*)$ gives $(aa)b = a(ab)$.

Property (12) is proved similarly on putting $1 = O$, $2 = B$,
$3 = A$, $4 = (1, b)$, $5 = (a, a(bb))$.

The class of alternative projective planes is a *proper* subclass
of the class of Veblen-Wedderburn projective planes; that is, the
minor theorem of Desargues might be satisfied *affinely* in a projective plane in which the minor theorem is not valid without the
affine restriction. To establish this we construct an example of
such a plane.

Let K denote the class of all expressions $x_0 + x_1 e_1 + x_2 e_2 + x_3 e_3$,
where x_0, x_1, x_2, x_3 are real numbers and e_1, e_2, e_3 are "units" with
the following multiplication table:

$\cdot$	e_1	e_2	e_3
e_1	-1	e_3	$-2e_2$
e_2	$-e_3$	-1	$3e_1$
e_3	$2e_2$	$-3e_1$	-1 ;

that is, the product $e_i e_j$ is the element in the i-th row and j-th
column of the table. Real numbers commute and associate multi-
plicatively with all of the units. With the aid of this table any two
elements of K can be shown to have a product that is also an element of K. The product depends on the order of the factors. If

$x = x_0 + x_1e_1 + x_2e_2 + x_3e_3$ and $y = y_0 + y_1e_1 + y_2e_2 + y_3e_3$, then, by definition, $x + y = x_0 + y_0 + (x_1 + y_1)e_1 + (x_2 + y_2)e_2 + (x_3 + y_3)e_3$.

It is obvious that the system $[K, +, \cdot]$ has properties (2), (3), (4), (5), (6), and (10) of Theorem VI.7.1, and property (9*) of Theorem VI.8.1.

Property (7). Let $a, b \in K$, $a \neq 0$; that is $a = a_0 + a_1e_1 + a_2e_2 + a_3e_3$ with not *all* of the real numbers a_0, a_1, a_2, a_3 zero, and $b = b_0 + b_1e_1 + b_2e_2 + b_3e_3$. A unique element y of K exists such that $ay = b$ if and only if a unique quadruple of real numbers y_0, y_1, y_2, y_3 exist such that

$$(a_0 + a_1e_1 + a_2e_2 + a_3e_3)(y_0 + y_1e_1 + y_2e_2 + y_3e_3)$$
$$= b_0 + b_1e_1 + b_2e_2 + b_3e_3.$$

Using the multiplication table to perform the indicated multiplication, and equating the coefficients of 1, e_1, e_2, e_3, respectively, yields the following equations in y_0, y_1, y_2, y_3:

$$a_0y_0 - a_1y_1 - a_2y_2 - a_3y_3 = b_0,$$
$$a_1y_0 + a_0y_1 - 3a_3y_2 + 3a_2y_3 = b_1,$$
$$a_2y_0 + 2a_3y_1 + a_0y_2 - 2a_1y_3 = b_2,$$
$$a_3y_0 - a_2y_1 + a_1y_2 + a_0y_3 = b_3.$$

The determinant of the coefficients is

$$\begin{vmatrix} a_0 & -a_1 & -a_2 & -a_3 \\ a_1 & a_0 & -3a_3 & 3a_2 \\ a_2 & 2a_3 & a_0 & -2a_1 \\ a_3 & -a_2 & a_1 & a_0 \end{vmatrix},$$

which, developed, is

$$a_0^4 + 2a_1^4 + 3a_2^4 + 6a_3^4 + 3a_0^2a_1^2 + 4a_0^2a_2^2 + 7a_0^2a_3^2 + 5a_1^2a_2^2 + 8a_1^2a_3^2 + 9a_2^2a_3^2.$$

This is not zero, since, by hypothesis, not all of the *real* numbers a_0, a_1, a_2, a_3 are zero, and only in that case does the above expression vanish.

Hence the system of equations has a unique solution y_0, y_1, y_2, y_3, and property (7) is established.

In a similar manner property (8) is shown to be valid in $[K, +, \cdot]$.

Property (9) of Theorem VI.7.1 is an immediate consequence of properties (7) and (9*). It follows that the projective plane constructed over set K [with $T(a, m, b)$ defined to be $a \cdot m + b$], in the manner described in Theorem VI.4.1, is a Veblen-Wedderburn plane.

But the plane is *not* alternative, since $(e_1e_1)e_3 = -e_3$, whereas $e_1(e_1e_3) = e_1(-2e_2) = -2e_1e_2 = -2e_3$. Now $-e_3 \neq -2e_3$, since if equality held, multiplying both sides by e_3 would yield $1 = 2$.

Remark. The relations $a(ab) = (aa)b$, $(ab)b = a(bb)$ are weak forms of the associativity of multiplication.

It may be shown that in a projective plane with the minor Desargues theorem valid, other weak forms of associativity of multiplication hold; for example, $a^{-1}(ab) = b$ and $(ab)b^{-1} = a$, where a^{-1} is the unique element of Γ such that $a^{-1}a = aa^{-1} = 1$, and b^{-1} is defined similarly. We shall use the first of these relations, as well as $(ab)^{-1} = b^{-1}a^{-1}$, in the next section.

VI.9. Desarguesian and Pappian Planes

We consider, first, projective planes in which the theorem of Desargues (Section VI.6) is (projectively) valid.

DEFINITIONS. A correspondence (point-to-point and line-to-line) of the elements of a projective plane Π with the elements of the same plane, which is one-to-one with respect to points, one-to-one with respect to lines, and preserves the incidence relation (that is, if P, P' are corresponding points, and q, q' are corresponding lines, then P is on q if and only if P' is on q'), is called a *collineation* of Π. It is clear that the set of all collineations of Π form a *group*.

The collineation group of a projective plane π is *quadruply transitive*, provided, corresponding to each two quadruples P, Q, R, S and P', Q', R', S' of points of Π (no three points of either quadruple being collinear), there is a collineation in which P and P', Q and Q', R and R', S and S' are corresponding points.

THEOREM VI.9.1.   *The Desargues theorem is valid in an alternative plane π if and only if multiplication is associative in at least one planar ternary ring of Π.*

Proof.   Let $1, 2, 3$ and $1', 2', 3'$ be the respective vertices of two triangles perspective with respect to point 4. Let lines $(1, 2)$ and $(1', 2')$ meet at point 5, lines $(1, 3)$ and $(1', 3')$ at point 6, and lines $(2, 3)$ and $(2', 3')$ at point 7.

Select the coordinate system with base points $A = 1'$, $B = 3'$, $O = 2$, and I [the intersection of lines $(1, 1')$ and $(2, 6)$ (Figure 47)], and let $[\Gamma, T]$ denote the ternary ring associated with it.

Then point 6 has coordinate (1), and point 1 has coordinates $(1, t), t \in \Gamma$. Let the coordinates of $2'$ be $(a, ab), a \neq 0, b \neq 0$, and $c = b^{-1}(b - t + 1)$.

Now point 4 is the intersection of line $(1, 1')$ (equation $x = 1$) with line $(2, 2')$ (equation $y = xb$), and so has coordinates $(1, b)$. Point 5 is the intersection of line $(1, 2)$ (equation $y = xt$) and line $(1', 2')$ (equation $x = a$), and so has coordinates (a, at). Line $(1, 6)$ (equation $y = x + t - 1$) meets line $(3', 4)$ (equation $y = b$) at point 3, which, consequently, has coordinates $(b - t + 1, b)$. Finally, line $(2', 3')$ (equation $y = ab$) meets line $(2, 3)$ at point 7. Since $2 = O$, line $(2, 3)$ has equation $y = xm$; since $3(b - t + 1, b)$ is on that line, $b = (b - t + 1)m$. Since $b - t + 1 \neq 0$ (why?), $m = (b - t + 1)^{-1}b$, and so [using the property $d^{-1}(dg) = g$] line $(2, 3)$ has equation $y = x[(b - t + 1)^{-1}b]$. Hence point 7 has coordinates $((ab)[(b - t + 1)^{-1}b]^{-1}, ab)$.

Line $(5, 6)$ has equation $y = x + at - a$, so the incidence of this line with point 7 (that is, the validity of the Desargues theorem) is equivalent to the relation

$$ab = (ab)[(b - t + 1)^{-1}b]^{-1} + at - a,$$

or to

$$a(b - t + 1) = (ab)[(b - t + 1)^{-1}b]^{-1}.$$

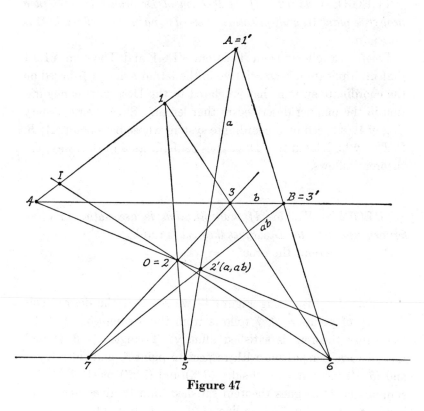

Figure 47

Using the relation $(de)^{-1} = e^{-1}d^{-1}$, we obtain

$$a(b - t + 1) = (ab)[b^{-1}(b - t + 1)].$$

Since $b - t + 1 = b[b^{-1}(b - t + 1)]$, the above relation is equivalent to

$$a\{b[b^{-1}(b - t + 1)]\} = (ab)[b^{-1}(b - t + 1)];$$

that is

$$a(bc) = (ab)c.$$

THEOREM VI.9.2.   *If the theorem of Desargues is valid in a
projective plane* Π, *multiplication in every ternary ring defined in* Π *is
associative.*

Proof.   It follows from Theorem VI.8.1 and Theorem VI.9.1
that multiplication is associative in the ternary ring of Π based on
the coordinate system that is related to the Desargues configuration in the manner described in that lemma. Since every ternary
ring of Π is based on a coordinate system whose base points A, B,
O, I may be related to a Desargues configuration in that way, the
theorem follows.

THEOREM VI.9.3.   *If multiplication is associative in every
ternary ring, then the Desargues theorem is valid.*

Proof.   We omit the proof.

Remark 1.   Inspecting Figure 47 we see that the desired col-
linearity of points *5*, *6*, *7* follows from the assumption that the
Desargues theorem is satisfied affinely. Triangles $(1, 3. 4)$ and
$(5, 7, 2')$ are perspective with respect to point *2*, and sides $(1, 4)$
and $(5, 2')$ meet at A, and sides $(3, 4)$ and $(7, 2')$ meet at B. Assuming the Desargues theorem satisfied affinely gives line $(5, 7)$
incident with point *6*, since line $(1, 3)$ meets line AB at point *6*.

Hence, the *affine satisfaction of the Desargues theorem implies the
associativity of one planar ternary ring of the plane and, consequently
(Theorem VI.9.3), the (projective) validity of the Desargues theorem.*

Remark 2.   What is referred to in this chapter as the affine
satisfaction of the Desargues theorem is called the third Desargues
property in Section V.6.

A projective plane is Pappian, provided the Pappus configuration (Section VI.6) is realizable in it; that is, if *1*, *3*, *5* are points
on one line, and *2*, *4*, *6* are points on another line, the intersection of
lines $(1, 2)$ and $(4, 5)$, of lines $(2, 3)$ and $(5, 6)$, and of lines $(3, 4)$
and $(6, 1)$ are all on the same line.

THEOREM VI.9.4. *A projective plane is Pappian if and only if, in every planar ternary ring of the plane, multiplication is commutative.*

Proof. Consider an arbitrary coordinate system with base points O, A, B, and complete the configuration as in Figure 48, labeling $A = 1$, $B = 8$, $O = 4$, $2 = (a, a)$, $6 = (b, b)$, $3 = (1, a)$ [point I is the intersection of lines $(1, 5)$ and $(2, 4)$], $a \neq b$.

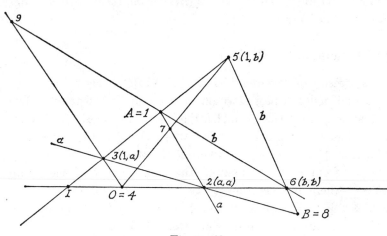

Figure 48

Labeling $5 = (1, b)$, line $(4, 5)$ has equation $y = xb$, so point 7 [the intersection of lines $(1, 2)$ and $(4, 5)$] has coordinates (a, ab). Lines $(2, 3)$ and $(5, 6)$ meet at point 8, and lines $(3, 4)$ and $(6, 1)$ meet at point 9, which [since line $(3, 4)$ has equation $y = xa$] has coordinates (b, ba).

Hence $ab = ba$ if and only if points 7 and 9 are on a line that is on B; that is, if and only if points 7, 8, 9 are on the same line.

We have shown (Theorem V.8.1) that the affine Pappus property implies the third Desargues property (that is, affine satisfaction of the Pappus theorem implies affine satisfaction of the Desargues theorem). But by a preceding remark, affine satisfaction of the Desargues theorem implies the projective validity of that theorem.

It follows that *in a projective plane the Pappus theorem implies the Desargues theorem.*

Combining this remark with Theorems VI.7.1, VI.9.2, and VI.9.4 gives the following important results.

THEOREM VI.9.5. *In a Pappian projective plane every (planar) ternary ring of the plane is a field.*

THEOREM VI.9.6. *The projective plane constructed over any given field is Pappian.*

VI.10. Concluding Remarks

As a result of Theorems VI.7.1 and VI.9.2, every ternary ring $[\Gamma, T]$ defined in a projective plane in which the theorem of Desargues is (projectively) valid has the following properties: $[\Gamma, +]$ is an abelian (commutative) group, where, by definition, $a + b = T(a, 1, b)$; $[\Gamma', \cdot]$ is a group, where Γ' denotes the set obtained by deleting from Γ the element 0 (the unit element of the group $[\Gamma, +]$) where, by definition, $a \cdot b = T(a, b, 0)$; and if $a, b, c \in \Gamma$, then $a(b + c) = ab + ac$ and $(a + b)c = ac + bc$. Only commutativity of multiplication is lacking for Γ to be a *field.*

Examples of non-Desarguesian projective planes show that the coordinate set of an arbitrary projective plane does not have such a rich algebraic structure. But it is a remarkable fact that *such planes are never planes of a three-dimensional projective geometry, for in every plane of a three-dimensional projective geometry the theorem of Desargues is valid.*

Axioms for a projective space of at least three dimensions (expressed in terms of a set S of abstract elements, called points, and certain subsets of S, called lines) are as follows:

$A_1.$ *If $A, B \in S$, $A \neq B$, there is exactly one line p such that A, $B \in p$.*

$A_2.$ *If $A, B, C \in S$, not all elements of any one line, and D, $E \in S$ $(D \neq E)$ such that $D \in line\ BC$, $E \in line\ AC$, there is a point F such that $F \in line\ AB$ and $F \in line\ DE$.*

DEFINITION. If $P, Q, R \in S$, not all elements of the same line, the set of all points of S contained in at least one of the lines PX, for all points X of the line QR, is called a *plane* (determined by P and line QR).

$A_3.$ *There are at least three pairwise distinct points on every line.*
$A_4.$ *There exists at least one line.*
$A_5.$ *Not all points are on the same line.*
$A_6.$ *Not all points are on the same plane.*

DEFINITION. If $P, Q, R, U \in S$, not all elements of the same plane, the set of all points of S contained in at least one of the lines PX, for all points X of the plane determined by Q and line RU, is called a *three-space.*

If a *closure axiom*, to the effect that all points of S are contained in a three-space, is assumed, the axiom system would define three-dimensional projective geometry. Such an axiom is not needed for our purpose.

Every plane of the projective space defined by axioms A_1 through A_6 is a *projective plane in the sense of Section VI.1* (with the incidence relation interpreted in the obvious way). (Let the reader show this.) Moreover, in each such plane the theorem of Desargues is valid, hence every ternary ring of the plane is a field *except (perhaps) for commutativity of multiplication.*

In proving the Desargues theorem we shall use many consequences of the axioms (for example, each two lines of a plane intersect; if a plane contains points A, B, it contains the line AB; a plane is uniquely determined by any three of its points that are not contained in the same line; the points common to each two distinct planes of a three-space form a line; and so on) without establishing them.

THEOREM VI.10.1. (DESARGUES). *Let Π denote a plane of the space defined by axioms A_1 through A_6. If A, B, C and A', B', C'*

are corresponding vertices of two triangles of Π *that are perspective with respect to a point, the triangles are perspective with respect to a line.*

Proof. We show first that if two triangles of a *three-space*, which are *not* co-planar, are perspective with respect to a point, they are perspective with respect to a line. Let P, Q, R and P', Q', R' be two such triangles (Figure 49).

Since points P, Q and P', Q' are in a plane (determined by two lines intersecting at O), lines PQ and $P'Q'$ intersect, and their common point L is in the plane determined by P, Q, R as well as in

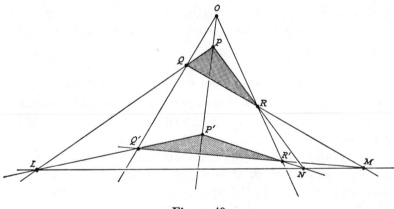

Figure 49

the plane determined by P', Q', R'. Similarly, points M and N, the intersections of lines QR and $Q'R'$ and of lines PR and $P'R'$, respectively, are in those two planes. Since these planes are in the three-space determined by O, P, Q, R, and they are distinct (since triangles P, Q, R and P', Q', R' are assumed non-coplanar), their intersection is a line. Consequently, the points L, M, N are on the same line.

Now let A, B, C and A', B', C' be vertices of two triangles in the plane Π, which are perspective with respect to the point O (Figure 50).

Let P, P' be any two distinct points of a line on O *that is not in*

plane π, $(P, P' \neq O)$. There is a unique three-space S_3 containing this line and the plane Π.

Points P, P' and A, A' are on two intersecting lines and hence are elements of the same plane. Hence lines PA and $P'A'$ intersect in a point A''. Similarly, lines PB and $P'B'$ intersect in B'', and lines PC and $P'C'$ intersect in C''. The triangle A'', B'', C'' is not in plane π, but it is perspective to triangle A, B, C from point P, and to triangle A', B', C' from point P'.

Consequently, by the result established in the first part of this proof, the lines $A''B''$, $B''C''$, $A''C''$ meet lines AB, BC, AC, re-

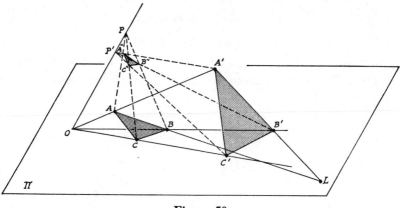

Figure 50

spectively, in three collinear points, say L, M, N, and those same lines meet lines $A'B'$, $B'C'$, $A'C'$, respectively, in three collinear points, say L', M', N'. But line $A''B''$ meets plane Π in exactly one point, therefore $L = L'$. Similarly, $M = M'$, $N = N'$, hence lines AB, BC, AC meet lines $A'B'$, $B'C'$, $A'C'$, respectively, in the three collinear points L, M, N, completing the proof of the theorem.

Remark. The theorem is trivial if the two perspective triangles have vertices in common.

To develop the analytic geometry of the projective plane it is convenient to employ homogeneous coordinates so that every point is represented by an ordered triple of elements of the coordinate set (which, as we have seen, is a field if the plane is Pappian), and triples (x_1, x_2, x_3), $(\rho x_1, \rho x_2, \rho x_3)$, $\rho \neq 0$, represent the same point. Triples $(x_1, x_2, 0)$ are points on line AB (in the familiar notation of this chapter), and since (by a change of coordinate systems) *any* line whatever of the plane can serve as this line, the "ideal locus" *is deprived of the special character it has in affine geometry.* Hence our transformations are not (as in affine geometry) restricted by the requirement of keeping the ideal line fixed, so the projective group of the plane is given by

$$\rho x_1' = a_{11}x_1 + a_{12}x_2 + a_{13}x_3,$$
$$\rho x_2' = a_{21}x_1 + a_{22}x_2 + a_{23}x_3,$$
$$\rho x_3' = a_{31}x_1 + a_{32}x_2 + a_{33}x_3,$$

$\rho \neq 0$, with the sole requirement that the determinant $|a_{ij}|$ $(i, j = 1, 2, 3)$ of the transformation be different from zero.

VII

Metric Postulates for the Euclidean Plane

Foreword

The reader might have noticed that in our study of the affine and projective planes no mention was made of *distance*—a notion that plays a fundamental role in that most important of all geometries, the euclidean. The reason for this is that distance is without significance in affine and projective geometries, since it is not an invariant of the groups that define those geometries. Two points with unit distance, for example, may be transformed by a transformation of either of those groups into two points with distance of one thousand units.

In euclidean geometry, however, the situation is quite different. The euclidean group does have the distance of two points as an invariant; that is, if points P, Q have distance d, every element of the euclidean group (that is, every translation and/or rotation) sends these points into points P', Q', respectively, with the *same* distance d. Hence distance is an object of study in euclidean geometry.

Though the euclidean group has other invariants than distance (for example, the angle one line makes with another), it is a remarkable fact that *every invariant can be expressed in terms of distances alone.* This suggests the desirability of a set of postulates for the euclidean plane in which distance is the sole primitive concept. Such an axiomatization of the euclidean plane is given in this chapter.

VII.1. Metric Space. Some Metric Properties of the Euclidean Plane

Let S denote any abstract set whose elements are called, for suggestiveness, points. If to each ordered pair p, q of elements of S a *non-negative* real number (denoted by pq and called, for suggestiveness, the metric or distance of p, q) is attached, such that

(1) $pq = 0$, if and only if $p = q$,

(2) $pq = qp$,

(3) for each three elements, p, q, r of S, $pq + qr \geqq pr$,

the resulting "space" is called a metric space M, over the *groundset* S, with *metric pq*.

Such spaces were introduced into mathematics by the French mathematician, Maurice Fréchet (1878–   ), nearly sixty years ago, and their metric and topological properties have been extensively investigated.

It follows immediately from the definition that $pq > 0$ if and only if $p \neq q$, and for each three points p, q, r of M, $|pq - qr| \leqq pr$.

Condition (3) for a metric space is called the *triangle inequality*, since it is satisfied by the lengths of the three sides of a euclidean triangle (Proposition 20, Book I, of the *Elements*), the equality sign holding if and only if the triangle is *degenerate* (that is, the three vertices are collinear). This condition is understood to be independent of the labeling of the three points involved, and, consequently, it implies the other two relations $pq + pr \geqq qr$, and $pr + qr \geqq pq$. A symmetric way of writing these three relations is $D(p, q, r) \leqq 0$, where

$$D(p, q, r) = \begin{vmatrix} 0 & 1 & 1 & 1 \\ 1 & 0 & pq^2 & pr^2 \\ 1 & pq^2 & 0 & qr^2 \\ 1 & pr^2 & qr^2 & 0 \end{vmatrix};$$

this bordered, symmetric, fourth-order determinant is easily written in factored form:

$$-(pq + qr + pr)(pq + qr - pr)(pq - qr + pr)(-pq + qr + pr).$$

(Let the reader show this.) We write

$$D(p, q) = \begin{vmatrix} 0 & 1 & 1 \\ 1 & 0 & pq^2 \\ 1 & pq^2 & 0 \end{vmatrix} = 2 \cdot pq^2.$$

A set N is called a subset of a metric space M, provided N is a subset of S, the groundset of M, and the distance of any two points p, q of N is the same as their distance in M. If N and N' are subsets of two metric spaces M and M', respectively, we say N is *congruent* to N' (written $N \approx N'$), provided there exists a one-to-one, *distance-preserving* correspondence between the points of N and the points of N'; that is, for every pair p, q of points of N, $pq = p'q'$, where p', q' are the points of N' that correspond, respectively, to points p, q of N and pq, $p'q'$ denote the distance of p, q in N and of p', q' in N', respectively. The spaces M, M' may be distinct or coincident. The latter relation is realized if and only if the groundset S of M is the same as the groundset S' of M', and each two points p, q have the same distance in M as in M'.

If two finite subsets $[p_1, p_2, \cdots, p_n]$, $[p_1', p_2', \cdots, p_n']$ of metric spaces are congruent and p_i, p_i' are corresponding points of the congruence $(i = 1, 2, \cdots, n)$, we write

$$p_1, p_2, \cdots, p_n \approx p_1', p_2', \cdots, p_n'.$$

In this case, $p_i p_j = p_i' p_j'$ $(i, j = 1, 2, \cdots, n)$.

A subset N of one metric space M is *congruently imbeddable* in a metric space M', provided there is a subset N' of M' such that $N \approx N'$. Then N is said to be *congruently contained* in M'.

THEOREM VII.1.1.   *Each set of three points of a metric space M is congruently contained in the euclidean plane.*

Proof.   Since for every non-negative number d the euclidean plane E_2 contains two points with distance d, the conclusion is clearly valid if the three given points p, q, r of M are not pairwise distinct.

If p, q, $r \in M$, pairwise distinct, let p', q' denote any two points of the plane such that p, $q \approx p'$, q' (that is, $p'q' = pq$). The circle $C(p'; pr)$, with center p' and radius pr, intersects (or is tangent to) the circle $C(q'; qr)$, with center q' and radius qr (since the sum of the radii is greater than or equal to the distance $p'q'$ of their centers). If r' denotes a point common to these two circles, clearly p, q, $r \approx p'$, q', r'.

COROLLARY VII.1.1.   *Each set of three points p, q, r of M is congruently contained in the euclidean straight line if and only if $D(p, q, r) = 0$.*

Let the reader supply the proof.

THEOREM VII.1.2.   *The euclidean plane is a metric space.*

Proof.   Conditions (1) and (2) are obviously satisfied if p, q, $r \in E_2$. If p, q, $r \in E_1$, the euclidean straight line, clearly one of the three numbers pq, qr, pr, is the sum of the other two numbers, and the triangle inequality is seen to be satisfied.

If p, q, $r \notin E_1$, then, by elementary trigonometry,

$$\cos p{:}q, r = (pq^2 + pr^2 - qr^2)/2pq \cdot pr,$$

where $p{:}q$, r denotes the angle at vertex p of the *non-degenerate* triangle of E_2 with vertices p, q, r. Then $-1 < \cos p{:}q$, $r < 1$ gives

$$-2pq \cdot pr < pq^2 + pr^2 - qr^2 < 2pq \cdot pr.$$

From the first inequality we obtain $pq + pr > qr$, and from the second we obtain $(-pq + pr + qr)(pq - pr + qr) > 0$. If both factors are negative, $qr < 0$, which is impossible; hence both factors are positive, and the result is $pr + qr > pq$, $pq + qr > pr$.

THEOREM VII.1.3. *If p, q, r, s are any four points of E_2, the determinant*

$$D(p, q, r, s) = \begin{vmatrix} 0 & 1 & 1 & 1 & 1 \\ 1 & 0 & pq^2 & pr^2 & ps^2 \\ 1 & pq^2 & 0 & qr^2 & qs^2 \\ 1 & pr^2 & qr^2 & 0 & rs^2 \\ 1 & ps^2 & qs^2 & rs^2 & 0 \end{vmatrix} = 0.$$

Proof. Let p, q, r, s be pairwise distinct points of E_2 (the theorem is obvious otherwise), and assume the labeling such that line (p, q) meets line (r, s) in point t (Figure 51).

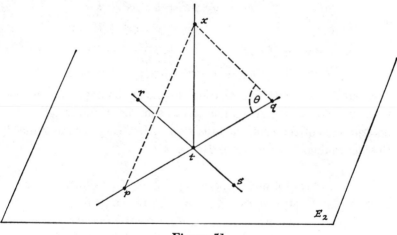

Figure 51

Let x denote any point of any plane containing line (p, q), $[x \notin \text{line } (p, q)]$. Applying the law of cosines to triangles p, q, x and q, t, x,

$$px^2 = pq^2 + qx^2 - 2pq \cdot qx \cos \theta,$$
$$xt^2 = qx^2 + qt^2 - 2qx \cdot qt \cos \theta.$$

Eliminating $\cos \theta$ from these two equalities, we see that constants a, b, c, d exist, not all zero, such that

(1) $a \cdot px^2 + b \cdot qx^2 + c \cdot tx^2 + d = 0, \quad a + b + c = 0.$

Applying this result to points r, s, t, x, constants a', b', c', d' exist, not all zero, such that

(2) $a' \cdot rx^2 + b' \cdot sx^2 + c' \cdot tx^2 + d' = 0, \quad a' + b' + c' = 0.$

Eliminating tx^2 from (1) and (2), we obtain constants e, f, g, h, k, not all zero, such that

$e \cdot px^2 + f \cdot qx^2 + g \cdot rx^2 + h \cdot sx^2 + k = 0, \quad e + f + g + h = 0.$

Let the reader show that these equalities hold also when three of the points p, q, r, s, are collinear.

By continuity, the first of these relations holds when $x = p, q, r$, s, in turn; that is,

$$f \cdot pq^2 + g \cdot pr^2 + h \cdot ps^2 + k = 0,$$
$$e \cdot pq^2 \qquad\qquad + g \cdot qr^2 + h \cdot qs^2 + k = 0,$$
$$e \cdot pr^2 + f \cdot qr^2 \qquad\qquad + h \cdot rs^2 + k = 0,$$
$$e \cdot ps^2 + f \cdot qs^2 + g \cdot rs^2 \qquad\qquad + k = 0.$$

Regarding the last five relations as linear, homogeneous equations in e, f, g, h, k having a non-trivial solution, it follows that the determinant of the coefficients is zero. After easy rearrangement this determinant is seen to be $D(p, q, r, s)$.

It is important for our purpose to point out *four* additional properties of E_2: First, if $p, r \in E_2$, $p \neq r$, there is a point q of E_2 such that $p \neq q \neq r$ and $pq + qr = pr$. Any point q *interior* to the line segment joining p, r has this property. Second, if $p, q \in E_2$, $p \neq q$, there is a point r of E_2 such that $pq + qr = pr$, $q \neq r$. Any point on the prolongation of the line segment joining p and q, *beyond* q, has the desired property. Third, the E_2 contains three points p, q, r such that $D(p, q, r) \neq 0$. An easy computation shows that the vertices p, q, r of any equilateral triangle of E_2 (for example, $pq = qr = pr = 1$) are such a triple.

The fourth property is known as *completeness*. It asserts that corresponding to any infinite sequence $p_1, p_2, \cdots, p_n, \cdots$ of points of E_2, such that $\lim p_i p_j = 0$, there is a point p of E_2 (necessarily unique) such that $\lim_{n \to \infty} pp_n = 0$. To consider this matter let us turn to the elementary analytic geometry of the euclidean plane, according to which a point is an ordered pair (x_1, x_2) of real numbers x_1, x_2, and the distance of two points (x_1, x_2), (y_1, y_2) is given by the expression $[(x_1 - y_1)^2 + (x_2 - y_2)^2]^{1/2}$.

Writing $p_n = (x_{1,n}, x_{2,n})$ $(n = 1, 2, \cdots)$ for the points of the infinite sequence, the condition $\lim_{ij \to \infty} p_i p_j = 0$ means that corresponding to every $\epsilon > 0$ there is a natural number N such that whenever each of the indices i, j exceeds N, the distance

$$[(x_{1i} - x_{1j})^2 + (x_{2i} - x_{2j})^2]^{1/2} < \epsilon.$$

It follows that each of the infinite sequences $\{x_{1n}\}$, $\{x_{2n}\}$ of real numbers x_{1n}, x_{2n} $(n = 1, 2, \cdots)$ has the property that for every $\epsilon > 0$ there is a natural number N such that for $i > N$ and $j > N$, $|x_{1i} - x_{1j}| < \epsilon$ and $|x_{2i} - x_{2j}| < \epsilon$. But this is precisely Cauchy's necessary and sufficient condition that real numbers x_1, x_2 exist such that $\lim_{n \to \infty} x_{1n} = x_1$ and $\lim_{n \to \infty} x_{2n} = x_2$.

Hence $\epsilon > 0$ implies the existence of a natural number N such that $n > N$ implies $|x_1 - x_{1n}| < \epsilon/\sqrt{2}$ and $|x_2 - x_{2n}| < \epsilon/\sqrt{2}$. Then for $n > N$, $[(x_1 - x_{1n})^2 + (x_2 - x_{2n})^2]^{1/2} < \epsilon$, so $\lim_{n \to \infty} pp_n = 0$, where $p = (x_1, x_2)$.

The first and second properties are stated by saying that the E_2 is *metrically convex* and *externally convex*. We have thus established that the euclidean plane E_2 has the following metric properties:

(i) The space E_2 is metric.

(ii) The space E_2 is metrically convex and externally convex.

(iii) The space E_2 is complete.

(iv) The space E_2 contains three points p, q, r such that $D(p, q, r) \neq 0$.

(v) The determinant $D(p, q, r, s)$ vanishes for every four points p, q, r, s of E_2.

We shall show that these metric properties are *characteristic* of the euclidean plane; that is, the euclidean plane possesses them, and *any* space that does is logically identical with the E_2. Hence these properties form a set of metric postulates for the euclidean plane; every euclidean property of the plane is a logical consequence of these five properties.

VII.2. A Set of Metric Postulates. The Space $\mathfrak{M}$

Let us consider an abstract set S whose elements are called points. To each pair p, q of points of S a non-negative real number pq is attached (called the distance of p, q) such that the resulting "space" $\mathfrak{M}$ satisfies the following postulates.

Postulate 1.   *The space $\mathfrak{M}$ is metric.*

Postulate 2.   *The space $\mathfrak{M}$ is metrically convex (that is, if p, $r \in \mathfrak{M}$, $p \neq r$, there exists an element q of $\mathfrak{M}$ such that $pq + qr = pr$ and $p \neq q \neq r$).*

Postulate 3.   *The space $\mathfrak{M}$ is externally convex (that is, if p, $q \in \mathfrak{M}$, $p \neq q$, there exists a point r of $\mathfrak{M}$ such that $pq + qr = pr$, and $q \neq r$).*

Postulate 4.   *The space $\mathfrak{M}$ is complete (that is, if $p_1, p_2, \cdots, p_n, \cdots$ is any infinite sequence of points of $\mathfrak{M}$ such that $\lim\limits_{ij \to \infty} p_i p_j = 0$, there exists a point p of $\mathfrak{M}$ such that $\lim\limits_{n \to \infty} p p_n = 0$).*

Postulate 5.   *The space $\mathfrak{M}$ contains three points a, b, c such that no one of the distances ab, bc, ac is equal to the sum of the other two; that is, $D(a, b, c) \neq 0$.*

Postulate 6.   *For every quadruple p, q, r, s of points of $\mathfrak{M}$, $D(p, q, r, s) = 0$.*

We write $\lim\limits_{n \to \infty} p_n = p$ (and say that the sequence $\{p_n\}$ has limit p) if and only if $\lim\limits_{n \to \infty} p p_n = 0$, where $p, p_1, p_2, \cdots, p_n, \cdots \in \mathfrak{M}$.

It was shown in the preceding section that the euclidean plane satisfies these postulates (and hence is a *model* for the system). We propose to show, in effect, that this postulational system is *categorical;* that is, any model of the system is isomorphic to the euclidean plane. As a result of this categoricity we may regard Postulates 1–6 as a set of (metric) postulates for euclidean plane geometry (metric, because the postulates are wholly and explicitly expressed in terms of the distance of point pairs).

We remark that the three points of Postulate 5 are necessarily pairwise distinct, and an infinite sequence $\{p_n\}$ of points of $\mathfrak{M}$ has *at most* one limit. Let the reader establish this.

VII.3. An Important Property of Space $\mathfrak{M}$

A point q of $\mathfrak{M}$ is *between* two points p, r of $\mathfrak{M}$, provided $pq + qr = pr$ and $p \neq q \neq r$. Three points of $\mathfrak{M}$ are called *linear*, provided they are congruently imbeddable in the straight line E_1. It follows that three pairwise distinct points of $\mathfrak{M}$ are linear if and only if one of the points is between the other two. By Corollary VII.1.1, three points p, q, r of $\mathfrak{M}$ are linear if and only if $D(p, q, r) = 0$.

THEOREM VII.3.1. *Each set of four points of $\mathfrak{M}$ is congruently imbeddable in the euclidean plane.*

Proof. According to Theorem VII.1.1, only quadruples of pairwise distinct points need be considered. Letting p, q, r, s denote such a quadruple of $\mathfrak{M}$, it is convenient to distinguish two cases.

Case 1. The quadruple p, q, r, s contains two linear triples. Assume the labeling such that p, q, r and p, q, s are linear triples. Then points p', q', r' and p'', q'', s'' of E_1 exist such that

$$(1) \qquad \begin{aligned} p, q, r &\approx p', q', r', \\ p, q, s &\approx p'', q'', s''. \end{aligned}$$

Since $p'q' = pq = p''q''$, the E_1 contains a unique point s' such that $p's' = p''s''$ and $q's' = q''s''$; then $p', q', s' \approx p'', q'', s''$, so

$$(2) \qquad p, q, s \approx p', q', s'.$$

It follows from congruences (1) and (2) that five of the six distances determined by the four points p', q', r', s' of E_1 equal the corresponding distances determined by the points p, q, r, s of $\mathfrak{M}$.

The theorem in Case 1 will be proved when the two sixth distances rs and $r's'$ are shown to be equal.

Let $D(p, q, r, s; x)$ denote the polynomial obtained on replacing the element rs^2 in the determinant $D(p, q, r, s)$ by the indeterminate x; that is,

$$D(p, q, r, s; x) = \begin{vmatrix} 0 & 1 & 1 & 1 & 1 \\ 1 & 0 & pq^2 & pr^2 & ps^2 \\ 1 & pq^2 & 0 & qr^2 & qs^2 \\ 1 & pr^2 & qr^2 & 0 & x \\ 1 & ps^2 & qs^2 & x & 0 \end{vmatrix}.$$

By congruences (1) and (2), it is clear that

$$D(p, q, r, s; x) \equiv D(p', q', r', s'; x),$$

and, consequently, the equations obtained by setting each of these polynomials equal to zero have the *same roots*. By Postulate 6, $x = rs^2$ is a root of $D(p, q, r, s; x) = 0$, and according to Theorem VII.1.3, $x = (r's')^2$ is a root of $D(p', q', r', s'; x) = 0$.

But since the triple p, q, r is linear, $D(p, q, r) = 0$, and it follows that

$$D(p, q, r, s; x) = -[Ax + B]^2/2pq^2,$$

where $A = 2pq^2$ and

$$B = \begin{vmatrix} 0 & 1 & 1 & 1 \\ 1 & 0 & pq^2 & pr^2 \\ 1 & pq^2 & 0 & qr^2 \\ 1 & ps^2 & qs^2 & 0 \end{vmatrix}.$$

Let the reader show this.

Hence $D(p, q, r, s; x) = 0$ *has exactly one root*, and so $rs^2 = r's'^2$; that is, $p, q, r, s \approx p', q', r', s'$, points of E_1.

Case 2.   The quadruple contains at most one linear triple.

Assume the labeling such that p, q, r and p, q, s are **non-linear** triples. By Theorem VII.1.1,

$$p, q, r \approx p', q', r',$$

$$p, q, s \approx p'', q'', s'',$$

where p', q', r' and p'', q'', s'' are points of E_2, and neither the first three of these points nor the last three are on straight lines.

Since $p'q' = pq = p''q''$, the plane contains two points s' and s^* such that p', q', $s' \approx p''$, q'', $s'' \approx p'$, q', s^*. Clearly, s' and s^* are reflections of each other in the line p', q' (Figure 52). [Since q, r, s or p, r, s may be linear, s' and/or s^* may be on line (q', r') or on line (p', r'), but our argument is independent of those possibilities.] The points s', s^* are *distinct*, otherwise they would coincide with a point of line p', q', which would imply that the triple p, q, s is linear, contrary to assumption.

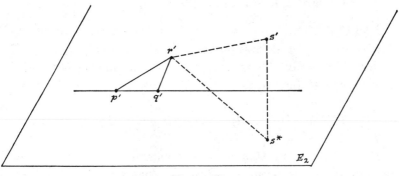

Figure 52

We shall show that $rs = r's'$ or $rs = r's^*$. In either event p, q, r, s are congruently imbeddable in E_2 (p, q, r, $s \approx p'$, q', r', s', if the first alternative holds, and p, q, r, $s \approx p'$, q', r', s^*, if the second alternative subsists).

As in the proof of Case 1,

$$D(p, q, r, s; x) \equiv D(p', q', r', s'; x) \equiv D(p', q', r', s^*; x).$$

Each of these polynomials is a quadratic $Ax^2 + Bx + C$, with $A = 2pq^2$, hence each has *exactly* two zeros. Now by Postulate 6, $x = rs^2$ is a zero, and by Theorem VII.1.3, $x = (r's')^2$ and $x = (r's^*)^2$ are also zeros. Since $r's' \neq r's^*$ [otherwise r' would lie on line (p', q') and p, q, r would be linear, contrary to assumption], it follows that $rs = r's'$ or $rs = r's^*$, and the theorem is proved.

COROLLARY VII.3.1.   *A quadruple of $\mathfrak{M}$ is linear (that is, congruently imbeddable in a straight line E_1) if and only if the quadruple contains at least two linear triples.*

By Theorem VII.3.1 every quadruple of $\mathfrak{M}$ is congruent with a quadruple of E_2, and each quadruple of E_2 is linear (that is, contained in a straight line) if and only if at least two of its triples are. Since linearity is obviously a congruence invariant, the corollary is established.

VII.4.   Straight Lines of Space $\mathfrak{M}$

DEFINITION.   *If p, q are any two distinct points of $\mathfrak{M}$, the straight line $L(p, q)$ determined by p, q is the set of all points x of $\mathfrak{M}$ such that p, q, x are linear (that is, congruently imbeddable in E_1).* Clearly, $p, q \in L(p, q)$.

THEOREM VII.4.1.   *If p, $q \in \mathfrak{M}$, $p \neq q$, $L(p, q)$ is congruently imbeddable in E_1.*

Proof.   Denote by p', q' any two points of E_1 such that $p'q' = pq$, and let x be any element of $L(p, q)$. Then p, q, $x \approx p''$, q'', x'', points of E_1, and since $p'q' = pq = p''q''$, E_1 contains exactly one point x' such that p', q', $x' \approx p''$, q'', x''. Hence, to each element x of $L(p, q)$ there corresponds a unique element x' of E_1, and p, q, $x \approx p'$, q', x'. To show that this correspondence is a congruence, let x', y' be the points of E_1, corresponding to points x, y of $L(p, q)$, respectively. We wish to show that $xy = x'y'$.

Now by Corollary VII.3.1,

$$p, q, x, y \approx p_1, q_1, x_1, y_1, \text{ points of } E_1,$$

and since $p_1q_1 = pq = p'q'$, E_1 contains exactly one point x^* and exactly one point y^* such that

$$p', q', x^*, y^* \approx p_1, q_1, x_1, y_1 \approx p, q, x, y.$$

Then p', q', $x^* \approx p$, q, $x \approx p'$, q', x' implies $x^* = x'$ and p', q', $y^* \approx p$, q, $y \approx p'$, q', y' implies $y^* = y'$ (since there is at most one point of E_1 with given distances from two points of E_1). Hence p, q, x, $y \approx p'$, q', x', y', and $xy = x'y'$, completing the proof of the theorem.

LEMMA VII.4.1. *If p, $q \in \mathfrak{M}$, $p \neq q$, the line $L(p, q)$ is deter-mined by any two of its distinct points.*

Proof. Let p^* be a point of $L(p, q)$, $p \neq p^* \neq q$, and consider $L(p^*, q)$. If $x \in L(p, q)$, then p, q, x are linear, and since $p^* \in L(p, q)$, then p, q, p^* are linear. Hence the quadruple p, q, p^*, x is congruently contained in E_1, and, consequently, all of its triples are linear. Then p^*, q, x is linear, so $x \in L(p^*, q)$. Thus, $L(p, q) \subset L(p^*, q)$, and in a similar manner it may be shown that $L(p^*, q) \subset L(p, q)$; that is, $L(p, q) = L(p^*, q)$. The same argument shows that $L(p^*, q) = L(p^*, q^*)$, where $q^* \in L(p^*, q)$, $p^* \neq q^* \neq q$, and, consequently, $L(p, q) = L(p^*, q^*)$, for any two distinct points p^*, q^* of $L(p, q)$.

By virtue of this lemma a line may be denoted by L without indicating a pair of its points that determines it.

COROLLARY VII.4.1. *Each two distinct points of $\mathfrak{M}$ are contained in exactly one line of $\mathfrak{M}$.*

LEMMA VII.4.2. *The metric (distance function) of a metric space is a continuous function of pointpairs of the space.*

Proof. Let p, q be any two points of $\mathfrak{M}$, and let $\{p_n\}$, $\{q_n\}$ be infinite sequences with $\lim p_n = p$, $\lim q_n = q$. Clearly,

$$|pq - p_n q_n| = |pq - pq_n + pq_n - p_n q_n|$$
$$\leq |pq - pq_n| + |pq_n - p_n q_n| \leq pp_n + qq_n.$$

Since $\lim_{n \to \infty} pp_n = 0$, $\lim_{n \to \infty} qq_n = 0$; then $\lim_{n \to \infty} |pq - p_n q_n| = 0$, and so $\lim p_n q_n = pq$.

A point p of $\mathfrak{M}$ is an *accumulation point* of a subset E of $\mathfrak{M}$, pro-vided that each $\epsilon > 0$ implies the existence of $q \in E$ such that

$0 < pq < \epsilon$. A subset E of $\mathfrak{M}$ is *closed*, provided that E contains all of its accumulation points.

THEOREM VII.4.2.   *Each line L of $\mathfrak{M}$ is metrically convex, externally convex, and closed.*

Proof.   Let L denote any line of $\mathfrak{M}$, and suppose p, $q \in L$, $p \neq q$. Since $\mathfrak{M}$ is metrically convex (Postulate 2), a point t of $\mathfrak{M}$ exists such that $pt + tq = pq, p \neq t \neq q$, and, consequently, p, q, t are linear. Since, by Lemma VII.4.1, $L = L(p, q)$, and the linearity of p, q, t implies that $t \in L(p, q)$, it follows that L is metrically convex.

Since $\mathfrak{M}$ is externally convex it contains a point r such that $pq + qr = pr, q \neq r$. Then p, q, r are linear, and $r \in L(p, q) = L$. Hence L is externally convex.

Let t be a point of $\mathfrak{M}$, and suppose $\{t_n\}$ is any infinite sequence of points of L with $\lim\limits_{n \to \infty} t_n = t$. If p, $q \in L$, $p \neq q$, then $L = L(p, q)$, and p, q, t_n are linear $(n = 1, 2, \cdots)$. Hence $D(p, q, t_n) = 0$ $(n = 1, 2, \cdots)$, and, consequently,

$$\lim_{n \to \infty} D(p, q, t_n) = 0.$$

Since the determinant $D(p, q, t_n)$ is a continuous function of its elements pt_n, qt_n, and since (Lemma VII.4.2) each of these elements is a continuous function,

$$\lim_{n \to \infty} D(p, q, t_n) = D(p, q, t).$$

Hence $D(p, q, t) = 0$, and $t \in L(p, q)$. It follows that the set L is closed.

THEOREM VII.4.3.   *Each line L of $\mathfrak{M}$ is congruent with the euclidean straight line E_1.*

Proof.   According to Theorem VII.4.1, E_1 contains a subset L' that is congruent to L. Since the properties of L established in Theorem VII.4.2 are clearly possessed by any set congruent to L (that is, those properties are *congruence invariants*), L' is metrically convex, externally convex, and closed (since $\mathfrak{M}$ is complete).

Suppose $t \in E_1$ and $t \notin L'$. Traversing E_1 from t to the right we encounter a first point of L', say q', or traversing E_1 from t to the left we encounter a first point of L', say p' (this is a consequence of L' being closed). If both alternatives hold ("or" is used here in the *inclusive* sense), then, since L' is metrically convex, it contains a point *between* p' and q', which contradicts the definitions of p' and q'. But it is impossible for exactly one of the alternatives to hold. Suppose the first, but not the second, subsists. Then no point of L' can lie to the right of q', for if there were such a point, the external convexity of L' would imply a point of L' to the left of q'. Such a point cannot be to the right of t (for then the definition of q' would be contradicted), nor can it lie to the left of t (for then the second alternative would hold).

Hence the assumption that there is a point of E_1 that does not belong to L' is untenable, so $L' = E_1$.

VII.5. Oriented Lines, Angles, and Triangles of Space $\mathfrak{M}$

DEFINITIONS. A line of $\mathfrak{M}$ is *oriented* by selecting an *ordered* pair p, q of its distinct points. The orientation given to L by the pointpair p, q is the *same* as the orientation given to L by the pointpair r, s, provided $D(p, q; r, s) = ps^2 + qr^2 - pr^2 - qs^2 > 0$, and the orientations are *opposite*, provided $D(p, q; r, s) < 0$ ($p, q, r, s \in L$, $p \neq q$, $r \neq s$).

Let $L_{p,q}$ denote a line oriented by the ordered pointpair p, q. We say it is *directed from p to q*.

Remark 1. Since $L \approx E_1$, it is easily seen that p, q, r, $s \in L$, $p \neq q$, $r \neq s$, implies $D(p, q; r, s) \neq 0$, and, consequently, each ordered pair of distinct points of L gives L either the same orientation as p, q (in this case we write $L_{p,q} = L_{r,s}$) or the opposite orientation.

Remark 2. Since $D(p, q; r, s) = -D(p, q; s, r)$, exactly one of the relations $L_{p,q} = L_{r,s}$, $L_{p,q} = L_{s,r}$ holds for each two ordered pairs of distinct points p, q and r, s of L. It follows that if p, q are any two distinct points of a line L, then $L_{p,q}$ or $L_{q,p}$ equals $L_{r,s}$ for each ordered pair of distinct points r, s of L.

DEFINITION.   Let $L_{p,q}$ and $L^*_{r,s}$ be any two directed lines of $\mathfrak{M}$. The *angle* made by these directed lines is the smallest non-negative value of θ such that

$$\cos \theta = D(p, q; r, s)/[D(p, q) \cdot D(r, s)]^{1/2}.$$

This definition requires justification. It must be shown that $|D(p, q; r, s)/[D(p, q) \cdot D(r, s)]^{1/2}| \leqq 1$, and that the right-hand member is invariant of a choice of the particular pointpairs p, q and r, s, so long as the orientations of the two lines are preserved.

Noting that

$$D(p, q; r, s) = \begin{vmatrix} 0 & 1 & 1 \\ 1 & pr^2 & ps^2 \\ 1 & qr^2 & qs^2 \end{vmatrix},$$

we write, similarly,

$$D(p, q, x; r, s, y) = \begin{vmatrix} 0 & 1 & 1 & 1 \\ 1 & pr^2 & ps^2 & py^2 \\ 1 & qr^2 & qs^2 & qy^2 \\ 1 & xr^2 & xs^2 & xy^2 \end{vmatrix},$$

and prove the following determinant theorem, which we shall find useful.

LEMMA VII.5.1.   If $a, b, c, d, p, q, r, s \in \mathfrak{M}$, $D(a, b; p, q) \cdot D(c, d; r, s) - D(a, b; r, s) \cdot D(c, d; p, q) = -D(a, b, c; p, q, r) - D(a, b, d; p, q, s) + D(a, b, c; p, q, s) + D(a, b, d; p, q, r)$.

Proof.   Let R denote the right-hand member of the equality asserted by the theorem. Writing the determinants symbolized and performing the indicated operations, we easily obtain

$$R = \begin{vmatrix} 0 & 1 & 1 & 0 \\ 1 & ap^2 & aq^2 & as^2 - ar^2 \\ 1 & bp^2 & bq^2 & bs^2 - br^2 \\ 0 & cp^2 - dp^2 & cq^2 - dq^2 & cs^2 - cr^2 - ds^2 + dr^2 \end{vmatrix}$$

$$= \begin{vmatrix} 1 & as^2 - ar^2 \\ 1 & bs^2 - br^2 \end{vmatrix} \times \begin{vmatrix} 1 & 1 \\ cp^2 - dp^2 & cq^2 - dq^2 \end{vmatrix}$$

$$+ [cs^2 - cr^2 - ds^2 + dr^2]D(a, b; p, q)$$

$$= -D(a, b; r, s) \cdot D(c, d; p, q) + D(c, d; r, s) \cdot D(a, b; p, q).$$

The reader is asked to verify the computations.

Let us write

$1 - D^2(p, q; r, s)/D(p, q) \cdot D(r, s)$

$$= [D(p, q) \cdot D(r, s) - D^2(p, q; r, s)]/D(p, q) \cdot D(r, s).$$

Applying the above determinant theorem (for $a = p, b = q, c = r,$ $d = s$) to the numerator of the right-hand member of this equality (noting that $D(p, q; p, q) = D(p, q)$, $D(p, q; r, s) = D(r, s; p, q)$, etc.) yields

$D(p, q) \cdot D(r, s) - D^2(p, q; r, s)$

$$= -D(p, q, r) - D(p, q, s) + 2D(p, q, r; p, q, s).$$

Since $D(p, q, r, s) = 0$, and

$(*)$ $D(p, q) \cdot D(p, q, r, s)$

$$= D(p, q, r) \cdot D(p, q, s) - D^2(p, q, r; p, q, s)$$

(let the reader verify this), we have

$\quad 1 - D^2(p, q; r, s)/D(p, q) \cdot D(r, s)$

$$= \frac{-D(p, q, r) - D(p, q, s) \pm 2[D(p, q, r) \cdot D(p, q, s)]^{1/2}}{D(p, q) \cdot D(r, s)}$$

$$= [\sqrt{-D(p, q, r)} \pm \sqrt{-D(p, q, s)}]^2/D(p, q) \cdot D(r, s).$$

Recalling that the D determinant of each three points of $\mathfrak{M}$ is negative or zero, and $D(p, q) > 0$, $D(r, s) > 0$, it follows that $1 - D^2(p, q; r, s)/D(p, q) \cdot D(r, s) \geqq 0$, and, consequently,

$$|D(p, q; r, s)/[D(p, q) \cdot D(r, s)]^{1/2}| \leqq 1.$$

Hence the angle θ defined above is a real angle, and $0 \leqq \theta \leqq \pi$.

To show that $D(p, q; r, s)/[D(p, q) \cdot D(r, s)]^{1/2}$ is invariant under any choice of points p, q, r, s that keeps the orientations of their respective lines unaltered, suppose $t \in L^*(r, s), s \neq t$.

Since $D(r, s, t, p) = 0$ and $D(r, s, t) = 0$, it follows from the relation

$$D(r, s) \cdot D(r, s, t, p) = D(r, s, t) \cdot D(r, s, p) - D^2(r, s, t; r, s, p)$$

that $D(r, s, t; r, s, p) = 0$. Expanding this determinant in a manner similar to that employed in $(*)$, we obtain

$$D(r, p; s, t) \cdot D(r, s) - D(r, p; r, s) \cdot D(r, s; s, t) = 0.$$

Another consequence of $D(r, s, t) = 0$ is $D^2(r, s; s, t) = D(r, s) \cdot D(s, t)$ [again the kind of expansion used in (∗) is employed], and we obtain

(†) $D^2(r, p; s, t) \cdot D(r, s) = D^2(r, p; r, s) \cdot D(s, t).$

Similarly, it can be shown that

(††) $D^2(r, s; p, q) \cdot D^2(r, p; s, t) = D^2(s, t; p, q) \cdot D^2(r, s; r, p).$

Dividing (††) by (†), we obtain

$$D^2(r, s; p, q)/D(r, s) = D^2(s, t; p, q)/D(s, t),$$

and hence

$$D^2(r, s; p, q)/D(r, s) \cdot D(p, q) = D^2(s, t; p, q)/D(s, t) \cdot D(p, q).$$

Now, if $L_{r,s} = L_{s,t}$, it may be shown that

$$D(r, s; p, q)/[D(r, s) \cdot D(p, q)]^{1/2}$$
$$= D(s, t; p, q)/[D(s, t) \cdot D(p, q)]^{1/2},$$

and so $\cos \theta$ is unaltered when point r of $L_{r,s}^*$ is replaced by any point t of that line, provided $L_{s,t}^*$ has the same orientation as $L_{r,s}^*$. It follows that $\cos \theta$ is invariant, in the manner described above, and the justification of its definition is complete.

Remark 3.   If $L_{p,q}$, $L_{p,r}$ are two oriented lines intersecting at point p,

$$\cos \theta = D(p, q; p, r)/[D(p, q) \cdot D(p, r)]^{1/2}$$
$$= (pq^2 + pr^2 - qr^2)/2pq \cdot pr.$$

Let p', q', r' be points of E_2 such that $p, q, r \approx p', q', r'$. Then the angle θ' made by the two oriented lines $L'_{p',q'}$, $L'_{p',r'}$ of E_2 is given by

$$\cos \theta' = [(p'q')^2 + (p'r')^2 - (q'r')^2]/2p'q' \cdot p'r'$$
$$= [pq^2 + pr^2 - qr^2]/2pq \cdot pr$$
$$= \cos \theta,$$

hence $\theta' = \theta$ (since each angle is between 0° and 180°).

DEFINITION.   If p, q, r are pairwise distinct points of $\mathfrak{M}$, the angles $\angle(L_{p,q}, L_{p,r})$, $\angle(L_{q,p}, L_{q,r})$, $\angle(L_{r,p}, L_{r,q})$ are called the *angles of the triple p, q, r*.

THEOREM VII.5.1.   *The sum of the angles of any triple p, q, r of pairwise distinct points of $\mathfrak{M}$ is* 180°.

Proof.   Let p', q', r' be points of E_2 such that p, q, $r \approx p'$, q', r'. Then, by Remark 3, each of the three angles of the triple p, q, r is equal to the corresponding angle in the euclidean triangle with vertices p', q', r', and the theorem follows.

DEFINITION.   If $p \in \mathfrak{M}$, to each point q of $\mathfrak{M}$, $p \neq q$, there corresponds a subset $R(p; q)$ of $\mathfrak{M}$ consisting of p, q and all points x of $\mathfrak{M}$ such that either x is between p and q, or q is between p and x. Call $R(p; q)$ the *ray with initial point p that contains q*.

Remark 4.   If p, $q \in \mathfrak{M}$, $p \neq q$, and p, $q \approx p'$, q' $(p', q' \in E_2)$, then, since $L(p, q) \approx E_1(p', q')$ (Theorem VII.4.3), it follows at once that $R(p; q) \approx r(p'; q')$, where $r(p'; q')$ is the *euclidean* ray with initial point p' that contains q' [that is, $r(p'; q')$ consists of p' and all points of $E_1(p', q')$ on the same side of p' as q'].

THEOREM VII.5.2.   *Let p, p_1, p_2, p_3 be pairwise distinct points of $\mathfrak{M}$. The figure formed by the three rays $R(p; p_i)(i = 1, 2, 3)$ is congruently imbeddable in E_2.*

Proof.   According to Theorem VII.3.1, E_2 contains four points p', p_1', p_2', p_3' such that

$$p, p_1, p_2, p_3 \approx p', p_1', p_2', p_3'.$$

The congruences $R(p; p_i) \approx r(p'; p_i')$ $(i = 1, 2, 3)$ are valid by Remark 4, and so if $x \in R(p; p_1) \cup R(p; p_2) \cup R(p; p_3)$, there is a unique point x' of $r(p'; p_1') \cup r(p'; p_2') \cup r(p'; p_3')$ corresponding to it by virtue of the appropriate one of those congruences.

Let x', y' correspond to x, y in this way. To fix the ideas we may suppose $x \in R(p; p_1)$ and $y \in R(p; p_2)$. Then $x' \in r(p'; p_1')$ and $y' \in r(p'; p_2')$ (Figure 53).

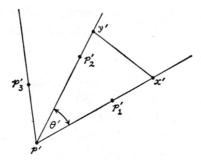

Figure 53

By Remark 3,

$$\angle\theta = \angle(L_{p,p_1}, L_{p,p_2}) = \angle[R(p;p_1), R(p;p_2)]$$
$$= \angle[r(p';p_1'), r(p';p_2')] = \theta',$$

and since

$$(x'y')^2 = (p'x')^2 + (p'y')^2 - 2(p'x')(p'y')\cos\theta'$$
$$= px^2 + py^2 - 2\cdot px\cdot py\cdot\cos\angle\theta$$
$$= xy^2$$

(and distances are non-negative), $xy = x'y'$, and the theorem is proved.

COROLLARY VII.5.1.   *The figure formed by the three rays $R(p;p_i)$ $(i = 1, 2, 3)$ is isogonally imbeddable in E_2.*

COROLLARY VII.5.2.   *The angles between three rays $R(p;p_i)$ $(i = 1, 2, 3)$ of $\mathfrak{M}$ (with common initial point p) satisfy the determinant relation $\Delta(p;p_1, p_2, p_3) = |\cos\angle[R(p;p_i), R(p;p_j)]| = 0$ $(i, j = 1, 2, 3)$.*

Proof.   By the congruent imbedding of

$$R(p;p_1) \cup R(p;p_2) \cup R(p;p_3) \text{ in } E_2, \Delta(p;p_1, p_2, p_3)$$
$$= \Delta(p';p_1', p_2', p_3') = 4\sin A\cdot\sin B\cdot\sin C\cdot\sin D,$$

where

$$A = (\tfrac{1}{2})(\theta_{12}' + \theta_{23}' + \theta_{13}'),$$
$$B = (\tfrac{1}{2})(\theta_{12}' + \theta_{23}' - \theta_{13}'),$$
$$C = (\tfrac{1}{2})(\theta_{12}' - \theta_{23}' + \theta_{13}'),$$
$$D = (\tfrac{1}{2})(-\theta_{12}' + \theta_{23}' + \theta_{13}'),$$

θ'_{ij} denoting $\angle[r(p'; p'_i), r(p'; p'_j)]$. Since one of the angles θ'_{ij} is the sum of the other two, or $\theta'_{12} + \theta'_{23} + \theta'_{13} = 360°$, the corollary follows.

COROLLARY VII.5.3. *If five points of $\mathfrak{M}$ contain a linear triple, they are congruently imbeddable in E_2.*

Proof. Let p, q, r be a linear triple of $\mathfrak{M}$, with $pq + qr = pr$. If $s, t \in \mathfrak{M}$, the five points p, q, r, s, t are contained in the sum of three rays with initial point p, and hence are congruently imbeddable in E_2.

DEFINITION. If p, $q \in \mathfrak{M}$, $p \neq q$, the *segment* (p, q) [denoted by seg (p, q)] consists of p, q and all points of $\mathfrak{M}$ between p and q.

Remark 5. If $p, q \in \mathfrak{M}$, $p \neq q$, and $p, q \approx p', q'$ $(p', q' \in E_2)$, seg (p, q) is congruent to the straight line segment (p', q'). This congruence is contained in the congruence $L(p, q) \approx E_1(p', q')$, which is insured by Theorem VII.4.3.

DEFINITION. If p_1, p_2, $p_3 \in \mathfrak{M}$, $p_1 \neq p_2 \neq p_3 \neq p_1$, the figure seg $(p_1, p_2) \cup$ seg $(p_2, p_3) \cup$ seg (p_1, p_3) is called a *triangle with vertices* p_1, p_2, p_3, denoted by $t(p_1, p_2, p_3)$. These segments are the *sides* of $t(p_1, p_2, p_3)$ and the angles of $t(p_1, p_2, p_3)$ are the angles of the triple p_1, p_2, p_3.

Remark 6. The angle-sum of any triangle of $\mathfrak{M}$ is equal to 180° (Theorem VII.5.2).

Remark 7. If p_1, p_2, p_3 are pairwise distinct elements of $\mathfrak{M}$ and $p_1, p_2, p_3 \approx p'_1, p'_2, p'_3$ $(p'_1, p'_2, p'_3 \in E_2)$, then

$$t(p_1, p_2, p_3) \approx t'(p'_1, p'_2, p'_3),$$

where $t'(p'_1, p'_2, p'_3)$ denotes the euclidean triangle with vertices p'_1, p'_2, p'_3.

Let the reader supply the proof of this remark.

VII.6. Metric Postulates for Euclidean Plane Geometry

We are now in a position to prove that Postulates 1–6, Section VII.2, *define* the euclidean plane. It has already been shown that they are all valid in the E_2 (Section VII.1). In order to attain our objective, it remains to select an existing categorical postulational system for euclidean plane geometry and to show that each postulate of that system is a logical consequence of Postulates 1–6.

We consider the following postulational system for euclidean plane geometry, given by H. G. Forder in his book, *The Foundations of Euclidean Geometry*, Dover, 1958).

Forder's system begins with seven assumptions concerning a *primitive* relation of three points p, q, r, denoted by $[pqr]$. These assumptions are called *axioms of order*.

O_1.   *If p, q, r are three points such that $[pqr]$ holds, p, q, r are pairwise distinct.*

O_2.   *If $[pqr]$ holds, then $[qrp]$ does not.*

DEFINITION.   *If p, q are distinct points, the line $L(p, q)$ consists of p, q and all points x such that $[xpq]$ or $[pxq]$ or $[p\acute{q}x]$ holds.*

O_3.   *If r, $s \in L(p, q)$ $r \neq s$, then $p \in L(r, s)$.*

O_4.   *If $p \neq q$, there is at least one point r with $[pqr]$.*

O_5.   *There are three points not on the same line.*

O_6.   *If p, q, r are non-collinear points, and s, t are distinct points such that $[qrs]$ and $[rtp]$, a point x exists such that $x \in L(s, t)$ and $[pxq]$.*

O_7.   *If p, q, r are non-collinear points, then, for every point x, distinct points y, z of the perimeter of the triple p, q, r exist such that x, y, z are collinear. (The perimeter of the triple p, q, r consists of all points w such that $[pwq]$ or $[qwr]$ or $[rwp]$, together with p, q, and r).*

To relate this to our system we *define* or *interpret* the relation $[pqr]$ to mean $pq + qr = pr$, $p \neq q \neq r$. Then O_1 and O_2 are obviously valid; the definition of line carries over to our definition; O_3 follows from Lemma VII.4.1; O_4 is a direct consequence of Postulate 3; and O_5 follows from Postulate 5.

To prove O_6 it follows from Theorem VII.5.2 that the figure $R(r;p) \cup R(r;s) \cup R(r;q)$ is congruent with the euclidean figure ray $(r';p') \cup$ ray $(r';s') \cup$ ray $(r';q')$, where p', q', r', s', t' are points of E_2 such that

$$p, q, r, s, t \approx p', q', r', s', t'$$

(Corollary VII.5.3.) This congruence between the two sets of three rays is readily extended to a congruence between the two sets obtained by adjoining seg (p, q) to the first and seg (p', q') to the second of these sets (Figure 54).

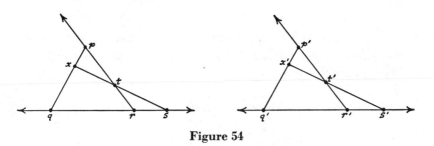

Figure 54

Now $L(s', t')$ intersects seg (p', q') in point x'. The point x that corresponds to x' in the congruence of the two figures is the desired point.

To prove O_7 let p, q, r, x be four points of $\mathfrak{M}$, p, q, r not on the same line, and let p', q', r', x' be points of E_2 such that $p, q, r, x \approx p', q', r', x'$. Clearly, x' is joined to at least one of the points p', q', r' by a line that has a point in common with the opposite side of the triangle. To fix the ideas, suppose $L(p', x')$ meets seg (q', r') at y', and let y be the point of $L(p, x)$ such that $p, x, y \approx p', x', y'$.

Since $p, q, r, x, y \approx \bar{p}, \bar{q}, \bar{r}, \bar{x}, \bar{y}$, points of E_2 (Corollary VII.5.3), and $p', q', r', x' \approx p, q, r, x \approx \bar{p}, \bar{q}, \bar{r}, \bar{x}$, there exists a *unique* point y^* of E_2 such that

$$p', q', r', x', y^* \approx \bar{p}, \bar{q}, \bar{r}, \bar{x}, \bar{y},$$

and so

$$p, q, r, x, y \approx p', q', r', x', y^*.$$

From $p', x', y' \approx p, x, y \approx p', x', y^*$, and the linearity of p', x', y', we conclude that $y^* = y'$. Since $q'y' + y'r' = q'r'$, then $qy + yr = qr$, and y is a point of seg (q, r). Hence x is collinear with two points (p and y) of the perimeter of the triple p, q, r. Postulate O_6 may be applied to p, q, x, y to show that x is collinear with two points of the perimeter of $t(p, q, r)$, neither of which is a vertex.

Following the seven axioms of order are seven axioms of *congruence*, a primitive relation of pointpairs. We *define* two pointpairs (p, q), (r, s) to be congruent (and write $p, q \approx r, s$), provided $pq = rs$.

C_1. *If $R(p; q)$ is a ray with initial point p, containing q, $p \neq q$, and $\rho > 0$ is any positive number, there is exactly one point r of $R(p; q)$ such that $pr = \rho$.*

This follows from Theorem VII.4.3.

C_2. *If $p, q \approx r, s \approx t, u$, then $p, q \approx t, u$.*

C_3. *If $[pqr]$, $[p'q'r']$ hold, and $p, q \approx p', q'$, while $q, r \approx q', r'$, then $p, r \approx p', r'$.*

C_4. *$p, q \approx q, p$.*

The proofs of C_2, C_3, and C_4 are obvious.

C_5. *If p, q, r are not collinear, p^*, q^*, r^* are not collinear, $[pqs]$, $[p^*q^*s^*]$ hold, $p, q \approx p^*, q^*$, and $q, r \approx q^*, r^*$, and $p, r \approx p^*, r^*$, and $p, s \approx p^*, s^*$, then $r, s \approx r^*, s^*$.*

From $[pqs]$, $[p^*q^*s^*]$, $pq = p^*q^*$, and $p, s \approx p^*, s^*$ we conclude $q, s \approx q^*, s^*$, and C_5 follows easily from Theorem VII.5.2.

C_6. *If $R(p; q)$ is a ray and θ is any given real number such that $0° < \theta° < 180°$, there are exactly two rays $R(p; v)$, $R(p; w)$ such that $\angle[R(p; q), R(p; v)] = \theta° = \angle[R(p; q), R(p; w)]$.*

Consider the line $L(p, q)$ that carries the ray $R(p; q)$, and let r be a point of $L(p, q)$ such that $rp + pq = rq, r \neq p \neq q$. There exists a point s of $\mathfrak{M}$, not on $L(p, q)$. Let p', q', r', s' be points of E_2 such that $p, q, r, s \approx p', q', r', s'$. Then triangle $t(q, r, s) \approx$ triangle $t(q', r', s')$. Let ray $(p'; t')$ be on the same side of $L(p', q')$ as s' and make angle $\theta°$ with ray $(p'; q')$. Now ray $(p'; t')$ intersects seg (q', s') or seg (r', s'). Suppose it meets seg (r', s'), and let w' denote the intersection (Figure 55). If w is the element of $\mathfrak{M}$ that corresponds to w' in the congruence of the two triangles, it is clear that $\angle[R(p; q), R(p; w)] = \angle[\text{ray } (p'; q'), \text{ray } (p'; w')] \triangleq \theta°$ (since $p, q \approx p', q'; p, w \approx p', w'; q, w \approx q', w')$.

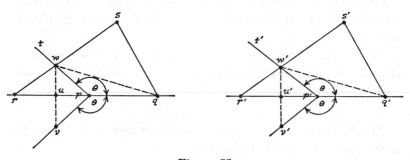

Figure 55

Let u' denote the foot of w' on $L(q', r')$. Assuming, as we may, that $qr > \max (qs, rs)$, it follows that $q'r' > \max (q'w', r'w')$, so $u' \in$ seg (q', r'). Let u correspond to u' in the congruence of the two triangles. Since $w'u' < w'x'$ for x' of seg (q', r'), $wu < wx$ for each point x of seg (q, r) [indeed, $wu < wx$ for each point x of $L(q, r)$]. We call u the *foot* of w on $L(q, r)$. Let v' be the point of $L(w', u')$ such that $w'u' + u'v' = w'v'$ and $w'u' = u'v'$, and let v be that point of $L(w, u)$ such that $w, u, v \approx w', u', v'$.

Clearly, $\angle[\text{ray } (p'; q'), \text{ray } (p'; v')] = \theta°$, ray $(p'; v') \neq$ ray $(p'; w')$, and from the congruence of p, q, w, u, v with five points of E_2 (Corollary VII.5.3), it follows readily that $p, q, w, u, v \approx p', q', w', u', v'$ and $\angle[R(p; q), R(p; v)] = \theta°$, $R(p; v) \neq R(p; w)$.

If ray $(p'; t')$ intersects seg (q', s'), the procedure is similar; and the proof of C_6 is complete.

C_7.   *Any two right angles are congruent.*

This is immediate from Theorem VII.5.2.

Axiom of Continuity.   *If all of the points of any line of $\mathfrak{M}$ fall into two subsets M and N such that*
(1) *each set contains at least two points,*
(2) *no point belongs to both M and N,*
(3) *if p is in one set and q, r are in the other, $[qpr]$ does not hold, then there is a point x such that $[mxn]$ holds for every $m \in M$ and $n \in N$, $m \neq x \neq n$.*
Proof.   Since each line of $\mathfrak{M}$ is congruent with a euclidean straight line, and the axiom is valid for such lines and is a congruence invariant, the axiom is valid in $\mathfrak{M}$.

The Parallel Axiom.   *There is not more than one line through a given point, parallel to a given line.*

According to Remark 6 of Section VII.5, the angle-sum of each triangle is equal to 180°, which (in the present context) is easily seen to imply the parallel axiom. See also the exercise below.
Thus, we have established the following result.

THEOREM VII.6.1.   *Postulates 1–6 form a postulational system for the euclidean plane.*

● EXERCISE

Prove the parallel axiom by showing that the figure formed by two mutually parallel lines of $\mathfrak{M}$ is congruently imbeddable in E_2, and that, consequently, two mutually parallel lines of $\mathfrak{M}$ are everywhere equidistant.

VII.7. Concluding Remarks

The task of showing that Postulates 1–6, Section VII.2, give a foundation of euclidean plane geometry would have been lightened considerably had we made use of results that we established elsewhere concerning similar postulates (Chap. V, L. M. Blumenthal, *Theory and Applications of Distance Geometry*, Clarendon, 1953). It was thought desirable, however, to make the developments of this book reasonably complete in themselves, so a new path leading to the proof of Theorem VII.6.1 was followed.

It is worth remarking that little else need be assumed in order to have a postulational basis for n-dimensional euclidean geometry, $n > 2$. Postulate 6 may be replaced by a *weaker* one; for example,

Postulate 6′. *If p, q, r, $s \in \mathfrak{M}$ such that $pq = qr = (\frac{1}{2})pr$, then $D(p, q, r, s) = 0$.*

Postulate 5 is altered as follows.

Postulate 5′. *There are $n + 1$ points whose D determinant does not vanish.*

The only additional postulate needed is the following one.

Postulate 7. *The D determinant of each set of $n + 2$ points vanishes.*

In the reference cited above it is shown that *any space $\mathfrak{M}$ for which Postulates 1, 2, 3, 4, 5′, 6′, and 7 are valid is congruent with n-dimensional euclidean space.*

Postulates for the
Non-euclidean Planes

Foreword

In Section I.7 we remarked that the term non-euclidean geometry was used first by Gauss to denote that geometry which arises on replacing Euclid's fifth postulate by its negation, and keeping unaltered all the remaining postulates that Euclid explicitly formulated or which enter implicitly into his development of geometry. In this sense there is but one non-euclidean geometry—Saccheri's geometry of the *acute-angle hypothesis*.

But from the time of Felix Klein a different nomenclature has been generally adopted. The geometry investigated by Sacheri, Gauss, Bolyai, and Lobachewsky is now known as hyperbolic geometry, whereas another geometry, in which *no two lines are mutually parallel* (and lines are finite in extent, but unbounded)—the so-called *elliptic geometry*—is also referred to as non-euclidean.

We shall round out our study of plane geometries by presenting (without proof) postulational systems for the hyperbolic and elliptic planes, as well as for the geometry of the surface of a sphere [sometimes called *double elliptic geometry*, since each two lines meet in *two* (diametral) points].

176

VIII.1. Poincaré's Model of the Hyperbolic Plane

We shall find it useful to employ a model for the hyperbolic plane introduced by the great French mathematician, Henri Poincaré (1854–1912).

A horizontal line L separates the euclidean plane into two parts: an "upper" half-plane and a "lower" half-plane (the line itself belongs to neither half-plane). The *points* of the hyperbolic plane are the points of the euclidean upper half-plane determined by L, and the *lines* of the hyperbolic plane consist of those euclidean (open) semi-circles, *with centers on L*, that lie in the upper half-plane of L, together with the upper half of *vertical* euclidean lines (Figure 56).

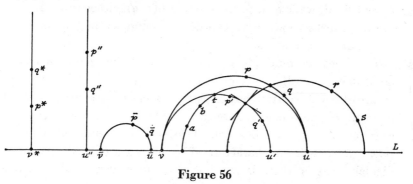

Figure 56

It is clear that corresponding to each two distinct points p, q of the hyperbolic plane there is a unique line containing both of them.

Let p, q be points of a hyperbolic line, and let their euclidean semi-circle intersect L in points u, v (remember that u and v are *not* points of the hyperbolic plane). If ρ is any arbitrarily chosen positive number, define the distance pq of the points p, q by the expression

$(*)$ $\qquad\qquad pq = (\rho/2) \ln (p, q; u, v),$

where $(p, q; u, v)$ denotes the cross ratio of the *ordered* quadruple of points p, q, u, v; that is, $(p, q; u, v) = (\overline{pu}/\overline{qu}) \cdot (\overline{qv}/\overline{pv})$, where the bar denotes the *euclidean distance* of points.

Since $(p, q; u, v) \geq 1$, it follows that $pq \geq 0$, and since $(p, q; u, v) = 1$ if and only if $p = q$, $pq = 0$ if and only if $p = q$. [In the event the hyperbolic line of p, q is a vertical euclidean half-line, the cross ratio $(p, q; u, v)$ becomes the simple ratio $\overline{pu}/\overline{qu}$ or $\overline{qv}/\overline{pv}$.]

Moreover,

$$qp = (\rho/2) \ln (q, p; v, u) = (\rho/2) \ln (p, q; u, v) = pq,$$

so distance is a *symmetric* function of pointpairs.

The hyperbolic plane, with distance defined in (∗), is said to have the "space constant ρ." We shall denote it by $H_{2,\rho}$.

It is convenient to call the points of L *ideal* points of $H_{2,\rho}$. If p is kept fixed, while q traverses the line $L(p, q)$, either toward u or toward v, distance pq increases without bound, so the lines of $H_{2,\rho}$ are infinite in extent.

Two lines are mutually *parallel*, provided they have an ideal point in common [for example, $L(p, q)$ and $L(p', q')$ of Figure 56] or are distinct vertical half-lines. It is clear that two distinct lines of $H_{2,\rho}$ may neither intersect nor be mutually parallel [for example, $L(p, q)$ and $L(\overline{p}, \overline{q})$]. Two such lines are called *non-intersectors;* they are characterized by having a common perpendicular (that is, there is one line perpendicular to each of two non-intersectors).

The model of the hyperbolic plane we are considering has the advantage of being *conformal* or *isogonal*. By this is meant that the angle between two hyperbolic lines is precisely the angle their counterparts make as figures in the euclidean plane. Thus the angle made by the lines $L(p', q')$ and $L(r, s)$ of Figure 56 is the (euclidean) angle made by the tangents to the two corresponding euclidean semi-circles at their point of intersection. Since their semi-circles are tangent at v, lines $L(p, q)$, $L(p', q')$ make an angle of 0°, and this is characteristic of mutually parallel lines.

Notice that through a point, not on a line, *two* lines may be drawn, each of which is parallel to the given line. Thus the lines

$L(p', q')$ and $L(a, b)$ are on point t, and each is parallel to line $L(p, q)$.

● EXERCISES

1. Construct the unique lines through a given point p that are parallel to a given line $L(q, r)$ not containing point p.
2. Construct the unique line through a point p that is perpendicular to a given line $L(q, r)$.
3. Construct the unique line that is perpendicular to each of two non-intersectors.

VIII.2. Some Metric Properties of the Hyperbolic Plane

If p, q, r are three points of $H_{2,\rho}$, consider the determinant

$$H(p, q, r) = \begin{vmatrix} 1 & \cosh(pq/\rho) & \cosh(pr/\rho) \\ \cosh(pq/\rho) & 1 & \cosh(qr/\rho) \\ \cosh(pr/\rho) & \cosh(qr/\rho) & 1 \end{vmatrix}.$$

Determinants of this kind play the same dominant role in hyperbolic geometry the D determinants do in euclidean geometry.

Multiplying the first column by $\cosh(pq/\rho)$ and $\cosh(pr/\rho)$, and subtracting the results from the second and third columns, respectively, we obtain

$$H(p, q, r) =$$
$$\begin{vmatrix} 1 - \cosh^2(pq/\rho) & \cosh(qr/\rho) - \cosh(pq/\rho)\cosh(pr/\rho) \\ \cosh(qr/\rho) - \cosh(pq/\rho)\cosh(pr/\rho) & 1 - \cosh^2(pr/\rho) \end{vmatrix}.$$

Since $1 - \cosh^2(pq/\rho) = -\sinh^2(pq/\rho)$, $1 - \cosh^2(pr/\rho) = -\sinh^2(pr/\rho)$, and application of the hyperbolic law of cosines to the triangle with vertices p, q, r gives

$$\cosh(qr/\rho) - \cosh(pq/\rho) \cdot \cosh(pr/\rho)$$
$$= -\sinh(pq/\rho) \cdot \sinh(pr/\rho) \cos \angle p\!:\!q, r,$$

(where $\angle p\!:\!q, r$ denotes the angle of the triangle at vertex p), substitution in the determinant yields

(†) $H(p, q, r) = \sinh^2(pq/\rho) \cdot \sinh^2(pr/\rho) \sin^2 \angle p\!:\!q, r.$

Hence the H determinant of each three points of $H_{2,\rho}$ is positive or zero. It is zero if and only if the three points are linear (that is, the sum of two of their three distances is equal to the third).

Developing $H(p, q, r)$ in another manner gives

(††) $H(p, q, r)$

$$= 4 \sinh (A/2) \cdot \sinh (B/2) \cdot \sinh (C/2) \cdot \sinh (D/2),$$

where
$$A = (pq + qr + pr)/\rho,$$
$$B = (pq + qr - pr)/\rho,$$
$$C = (pq - qr + pr)/\rho,$$
$$D = (-pq + qr + pr)/\rho.$$

From $H(p, q, r) \geqq 0$ it readily follows that $B \geq 0$, $C \geq 0$, $D \geq 0$, so points p, q, r satisfy the triangle inequality. Hence the hyperbolic plane $H_{2,\rho}$ is a *metric space*.

If p, q, r, s are any four points of $H_{2,\rho}$, we obtain, by a procedure similar to that employed to obtain (†)

$$H(p, q, r, s) = -\sinh^2 (pq/\rho) \sinh^2 (pr/\rho) \sinh^2 (ps/\rho) \cdot \Delta,$$

where
$$\Delta = \begin{vmatrix} 1 & \cos \angle p{:}q, r & \cos \angle p{:}q, s \\ \cos \angle p{:}q, r & 1 & \cos \angle p{:}r, s \\ \cos \angle p{:}q, s & \cos \angle p{:}r, s & 1 \end{vmatrix}.$$

The three angles appearing in Δ are made by the euclidean lines that are the tangents at p to the semi-circles that are the hyperbolic lines $L(p, q)$, $L(p, r)$, $L(p, s)$. It follows that *either* one of these angles is the sum of the other two, *or* the sum of the three angles is 360°. In either case, $\Delta = 0$ (see the proof of Corollary VII.5.2), and we have the following result.

THEOREM VIII.2.1.   *The determinant $H(p, q, r, s)$ of any four points p, q, r, s of $H_{2,\rho}$ vanishes.*

If p, q, r are points of a hyperbolic line, encountered in that order, $pq = (\rho/2) \ln (p, q; u, v)$, $qr = (\rho/2) \ln (q, r; u, v)$, $pr = (\rho/2) \ln (p, r; u, v)$, and

$$pq + qr = (\rho/2) \ln (p, q; u, v)(q, r; u, v)$$
$$= (\rho/2) \ln [(\overline{pu} \cdot \overline{qv})/\overline{qu} \cdot \overline{pv}][(\overline{qu} \cdot \overline{rv})/\overline{ru} \cdot \overline{qv}]$$
$$= (\rho/2) \ln [(\overline{pu} \cdot \overline{rv})/\overline{pv} \cdot \overline{ru}]$$
$$= (\rho/2) \ln (p, r; u, v)$$
$$= pr.$$

It follows that if three points are on a line, the sum of two of the three distances is equal to the third; the converse is easily seen to be valid also. From the expansion (††) of the determinant $H(p, q, r)$ we conclude that p, q, r are points of a hyperbolic line if and only if $H(p, q, r) = 0$. Another important conclusion is stated in the following theorem.

THEOREM VIII.2.2.   *The hyperbolic line is congruent with the euclidean straight line.*
We omit the proof.

It is clear from the foregoing that the hyperbolic plane is metrically convex, externally convex, and (in view of the model) complete.

VIII.3. Postulates for the Hyperbolic Plane

Consider an abstract set Σ, whose elements are called points, and a positive number ρ, which conform to the following postulates.

Postulate 1.   *Σ is a metric space.*
Postulate 2.   *The space Σ is metrically convex.*
Postulate 3.   *The space Σ is externally convex.*
Postulate 4.   *The space Σ is complete.*
Postulate 5.   *Three points a, b, c of Σ exist such that no one of the three distances ab, bc, ac equals the sum of the other two.*
Postulate 6.   *For every four points p_1, p_2, p_3, p_4 of Σ the determinant $H(p_1, p_2, p_3, p_4)$ vanishes.*

It can be proved that Σ is logically identical with the hyperbolic plane, with space constant ρ.

If $p, q \in \Sigma$, $p \neq q$, the line $L(p, q)$ is defined to be the set of all points x of Σ such that $H(p, q, x) = 0$, and $L(p, q)$ is shown to be congruent with a hyperbolic line.

It is instructive to compare the postulates for the hyperbolic plane given above with the postulates for the euclidean plane given in Section VII.2. Each set consists of six postulates, and the first five postulates of the two systems are identical. The two postulational systems differ only in their last postulates. For the E_2 it is assumed that every four points have their D determinant equal to zero, while for the $H_{2,\rho}$ it is supposed that the H determinant of every point-quadruple vanishes. All of the many differences between euclidean and hyperbolic geometry arise from the differences in the properties of those two determinants!

VIII.4. Two-dimensional Spherical Geometry

A model for two-dimensional spherical geometry is the surface of a sphere, with the distance of two points defined to be the length of the shorter arc of a great circle joining them. If the radius of the sphere is r $(r > 0)$, we denote the surface by $S_{2,r}$.

The points of $S_{2,r}$ are ordered triples (x_1, x_2, x_3) of real numbers, such that $x_1^2 + x_2^2 + x_3^2 = r^2$, and the distance of two points $x = (x_1, x_2, x_3)$ and $y = (y_1, y_2, y_3)$ is defined to be the *smallest non-negative* number xy such that

$$\cos (xy/r) = (x_1y_1 + x_2y_2 + x_3y_3)/r^2.$$

Hence $0 \leq xy \leq \pi r$.

It is clear that if $x = y$, then $xy = 0$. If, on the other hand, $xy = 0$, then $x_1y_1 + x_2y_2 + x_3y_3 = r^2$. Since

$$(x_1 - y_1)^2 + (x_2 - y_2)^2 + (x_3 - y_3)^2$$
$$= 2r^2 - 2(x_1y_1 + x_2y_2 + x_3y_3) = 0,$$

it follows from $xy = 0$ that $x_i = y_i$ $(i = 1, 2, 3)$, so $x = y$. Since the symmetry of the distance is obvious, it remains to establish the triangle inequality in order to conclude that $S_{2,r}$ is a metric space.

If $x = (x_1, x_2, x_3)$, $y = (y_1, y_2, y_3)$, and $z = (z_1, z_2, z_3)$ are any three points of $S_{2,r}$, we have

$$\begin{vmatrix} x_1 & x_2 & x_3 \\ y_1 & y_2 & y_3 \\ z_1 & z_2 & z_3 \end{vmatrix}^2 = \begin{vmatrix} x_1 & x_2 & x_3 \\ y_1 & y_2 & y_3 \\ z_1 & z_2 & z_3 \end{vmatrix} \cdot \begin{vmatrix} x_1 & y_1 & z_1 \\ x_2 & y_2 & z_2 \\ x_3 & y_3 & z_3 \end{vmatrix} = \begin{vmatrix} \Sigma x_i^2 & \Sigma x_i y_i & \Sigma x_i z_i \\ \Sigma y_i x_i & \Sigma y_i^2 & \Sigma y_i z_i \\ \Sigma z_i x_i & \Sigma z_i y_i & \Sigma z_i^2 \end{vmatrix},$$

where each summation is taken for $i = 1, 2, 3$.

Since $\Sigma x_i^2 = \Sigma y_i^2 = \Sigma z_i^2 = r^2$, the last determinant is seen to equal $r^6 \cdot \Delta(x, y, z)$, where

$$\Delta(x, y, z) = \begin{vmatrix} 1 & \cos(xy/r) & \cos(xz/r) \\ \cos(yx/r) & 1 & \cos(yz/r) \\ \cos(zx/r) & \cos(zy/r) & 1 \end{vmatrix},$$

and, consequently, *for each three points x, y, z of $S_{2,r}$ the determinant* $\Delta(x, y, z) \geqq 0$.

We have already noted (see proof of Corollary VII.5.2) that

$$\Delta(x, y, z) = 4 \sin(A/2) \cdot \sin(B/2) \cdot \sin(C/2) \cdot \sin(D/2),$$

where

$$A = (xy + yz + xz)/r,$$
$$B = (xy + yz - xz)/r,$$
$$C = (xy - yz + xz)/r,$$
$$D = (-xy + yz + xz)/r.$$

Since each of the angles xy/r, yz/r, xz/r is non-negative and at most π, it is easily seen that $\Delta(x, y, z) \geqq 0$ implies $B \geq 0$, $C \geq 0$, $D \geq 0$, so the distances xy, yz, xy satisfy the triangle inequality. We have proved the following theorem.

THEOREM VIII.4.1. *The space $S_{2,r}$ is metric.*

Now let p_1, p_2, p_3, p_4 be elements of $S_{2,r}$, and consider the determinant

$$\Delta(p_1, p_2, p_3, p_4) = |\cos(p_i p_j/r)| \qquad (i, j = 1, 2, 3, 4).$$

Multiplying the first column of the determinant by $\cos (p_1p_i/r)$ and subtracting from the i-th column, for $i = 2, 3, 4$, yields

$$\Delta(p_1, \cdots p_4) =$$

$$\begin{vmatrix} \sin^2 A_{12} & \cos A_{23} - \cos A_{12} \cos A_{13} & \cos A_{24} - \cos A_{12} \cos A_{14} \\ \cos A_{23} - \cos A_{12} \cos A_{13} & \sin^2 A_{13} & \cos A_{34} - \cos A_{19} \cos A_{14} \\ \cos A_{24} - \cos A_{12} \cos A_{14} & \cos A_{34} - \cos A_{14} \cos A_{13} & \sin^2 A_{14} \end{vmatrix}.$$

where $A_{ij} = p_ip_j/r \ (i, j = 1, 2, 3, 4; i \neq j)$.

Using the law of cosines of spherical trigonometry gives

$$(\S) \quad \Delta(p_1, \cdots, p_4) =$$

$$\sin^2 A_{12} \sin^2 A_{13} \sin^2 A_{14} \cdot \begin{vmatrix} 1 & \cos \angle 1{:}23 & \cos \angle 1{:}2, 4 \\ \cos \angle 1{:}2, 3 & 1 & \cos \angle 1{:}3, 4 \\ \cos \angle 1{:}24 & \cos \angle 1{:}34 & 1 \end{vmatrix},$$

where $\angle 1{:}2, 3$, $\angle 1{:}2, 4$, $\angle 1{:}3, 4$ are the three angles made by the three tangents to the great circles $C(p_1, p_2)$, $C(p_1, p_3)$, $C(p_1, p_4)$ at their common point p_1. These three tangents lie in the euclidean plane that is tangent to $S_{2,r}$ at p_1, and, consequently the determinant in $(\S)$ vanishes. This establishes the following important result.

THEOREM VIII.4.2. *The determinant* $\Delta(p_1, p_2, p_3, p_4)$ *vanishes for each four points* p_1, p_2, p_3, p_4 *of* $S_{2,r}$.

It is clear that $S_{2,r}$ is metrically convex and complete, but it is *not* externally convex; if $p, q \in S_{2,r}$ such that $pq = \pi r$, there is *no* point t of $S_{2,r}$ such that $pq + qt = pt$, $q \neq t$, since no two points of $S_{2,r}$ have distance exceeding πr. Instead of being externally convex, $S_{2,r}$ is *diameterized;* that is, corresponding to each point p of $S_{2,r}$ there is at least one point p^* of $S_{2,r}$ such that $pp^* = \pi r$.

VIII.5. Postulates for Two-dimensional Spherical Space

Let Σ denote any abstract set, and r any positive number. We make the following assumptions.

Postulate 1. Σ *is a metric space.*

Postulate 2. *The space Σ is metrically convex.*

Postulate 3. *The space Σ is diameterized; that is, $p \in \Sigma$ implies the existence of $p^* \in \Sigma$ such that $pp^* = \pi r$.*

Postulate 4. *The space Σ is complete.*

Postulate 5. *If $p, q \in \Sigma$, then $pq \leq \pi r$.*

Postulate 6. Σ *contains three points p_1, p_2, p_3 such that $\Delta(p_1, p_2, p_3) \neq 0$.*

Postulate 7. *The Δ determinant of each point quadruple of Σ vanishes.*

It may be shown that Σ is logically identical with $S_{2,r}$. Note that in spherical space the determinant Δ plays the dominant role.

VIII.6. The Elliptic Plane $\mathcal{E}_{2,r}$

A model for the elliptic plane is obtained by re-defining distance in the spherical space $S_{2,r}$. If $p, q \in S_{2,r}$, put $pq =$ spher dist (p, q) in the event that the spherical distance is less than or equal to $\pi r/2$, and define $pq = \pi r -$ spher dist (p, q) otherwise. Hence spherical distances of not more than a quadrant are unaltered, whereas those exceeding a quadrant are *replaced by their supplements*. This has the effect of giving spherically diametral points the distance zero, and, consequently, such pointpairs are *identified*. This is permissible, since each point q of $S_{2,r}$ has the same *elliptic* distances from each member of any pair of diametral points. Clearly the maximum distance of two points of $\mathcal{E}_{2,r}$ is $\pi r/2$.

The elliptic plane has certain peculiarities that make its study in the manner employed for the euclidean, hyperbolic, and spherical spaces quite difficult. Instead of one determinant, as in the previous spaces, the elliptic plane makes use of the *class of all determinants* obtained from $|\cos (p_i p_j/r)|$ $(i, j = 1, 2, \cdots, n)$ by means of a *symmetric alteration of the signs of its elements;* that is, it employs determinants

$$|\epsilon_{ij} \cos (p_i p_j/r)|. \quad \epsilon_{ij} = \epsilon_{ji} = \pm 1, \quad \epsilon_{ii} = 1 \quad (i, j = 1, 2, \cdots, n).$$

THEOREM VIII.6.1.   *The elliptic plane* $\mathcal{E}_{2,r}$ *is a metric space.*

Proof. If $p, q \in \mathcal{E}_{2,r}$, it is clear from the model that $pq \geqq 0$ and that $pq = 0$ if and only if $p = q$ (remember that diametral point-pairs of $S_{2,r}$ are identified in the $\mathcal{E}_{2,r}$); it is equally clear that $pq = qp$. It remains to prove the triangle inequality.

Let p, q, s be any three points of $\mathcal{E}_{2,r}$. Then either (i) all three of the elliptic distances pq, qs, ps are also the spherical distances of p, q, s (regarded as points of $S_{2,r}$), or (ii) it may be assumed that pq and qs are spherical distances, while ps is *not*. In the latter case, $ps = \pi r -$ spher dist (p, s), so cos $(ps/r) = -\cos (\widehat{ps}/r)$, where $\widehat{ps}$ denotes spher dist (p, s). Hence, from Section VIII.4 we may conclude that either (i) $\Delta(p, q, s) \geqq 0$, or (ii) $\Delta^*(p, q, s) \geqq 0$, where $\Delta^*(p, q, s)$ is obtained from $\Delta(p, q, s,)$ by prefixing a minus sign to the element in the second row and third column of $\Delta(p, q, s)$ as well as to the element in the third row and second column.

If $\Delta(p, q, s) \geqq 0$, it follows, as in Section VIII.4, that the distances pq, qs, ps satisfy the triangle inequality. Now it is easily seen that

$$\Delta(p, q, s) = \Delta^*(p, q, s) + 4 \cos (pq/r) \cos (qs/r) \cos (ps/r),$$

and since each factor in the second summand is non-negative (every distance in $\mathcal{E}_{2,r}$ is less than or equal to $\pi r/2$), $\Delta^*(p, q, s) \geqq 0$ implies $\Delta(p, q, s) \geqq 0$, so pq, qs, ps satisfy the triangle inequality in this case also.

THEOREM VIII.6.2.   *If* $p_1, p_2, p_3, p_4 \in \mathcal{E}_{2,r}$, *there exists a symmetric square matrix* (ϵ_{ij}), $\epsilon_{ij} = \epsilon_{ji} = \pm 1$, $\epsilon_{ii} = 1$ $(i, j = 1, 2, 3, 4)$ *such that the determinant* $|\epsilon_{ij} \cos (p_i p_j/r)|$ $(i, j = 1, 2, 3, 4)$ *vanishes.*

Proof. If $p_1, p_2, p_3, p_4 \in \mathcal{E}_{2,r}$, there are four points p_1', p_2', p_3', p_4' of $S_{2,r}$ such that $p_i'p_j'$ equals $p_i p_j$ or $\pi r - p_i p_j$, and, consequently, cos $(p_i'p_j'/r) = \epsilon_{ij} \cos (p_i p_j/r)$, $\epsilon_{ij} = \epsilon_{ji} = \pm 1$, $\epsilon_{ii} = 1$ $(i, j = 1, 2, 3, 4)$.

Since, by Theorem VIII.4.2, $|\cos (p_i'p_j'/r)|$ $(i, j = 1, 2, 3, 4)$ vanishes, so does $|\epsilon_{ij} \cos (p_i p_j/r)|$ $(i, j = 1, 2, 3, 4)$, and the theorem is proved.

VIII.7. Postulates for the Elliptic Plane

Let Σ denote an abstract set, whose elements are called points, and let r denote a given positive number. We make the following assumptions.

Postulate 1. Σ *is a metric space.*

Postulate 2. *The space Σ is metrically convex.*

Postulate 3. *If p, $q \in \Sigma$, $pq \leqq \pi r/2$.*

Postulate 4. *The space Σ is complete.*

Postulate 5. *If p, $q \in \Sigma$, $0 < pq < \pi r/2$, then Σ contains points p', q' such that $pq + qp' = pp'$, $qp + pq' = qq'$, and $pp' = qq' = \pi r/2$.*

Postulate 6. *There are three points u, v, w of Σ such that $\Delta^*(u, v, w) > 0$.*

Postulate 7. *For each four points p_1, p_2, p_3, p_4 of Σ there is a matrix (ϵ_{ij}), $\epsilon_{ij} = \epsilon_{ji} = \pm 1$, $\epsilon_{ii} = 1$ $(i, j = 1, 2, 3, 4)$ such that the determinant $|\epsilon_{ij} \cos (p_i p_j/r)|$ $(i, j = 1, 2, 3, 4)$ vanishes, and all third-order principal minors are non-negative.*

It may be shown that Postulates 1–7 form a postulational system for the elliptic plane with space constant r.

For the most recent investigation of the axiomatics of elliptic space of n-dimensions the reader is referred to the author's "New metric postulates for elliptic n-space," contained in *The Axiomatic Method, Studies in Logic and the Foundations of Mathematics,* North-Holland Publishing Company, Amsterdam, 1959.

Index

Accumulation point, 161–162
Acute-angle hypothesis, 10, 13, 15, 17
Affine: plane, 54, 55; defined by
  ternary ring, 65
  property, 105
  satisfaction of a theorem, 128
Aleph nought, $\aleph_0$, 27, 28, 29
Alternative plane, 134
Aristotle, 6, 7, 30, 37
Axiom, 2
  of Archimedes, 52
  of choice, 29, 30
  of completeness, 52
  of congruence, 52, 172
  of continuity, 52, 174
  of order, 51, 170

Beltrami, E., 15, 17, 41
Bernays, P., 42, 51, 52
Between, 4, 52, 157
Blumenthal, L. M., 175
Bolyai, J., 9, 11, 13, 14, 15, 53, 54, 176
Bolyai, W., 12, 13
Brouwer, L. E. J., 37
Bruck, R. H., vi

Cantor, G., 25
Cantor–Bernstein theorem, 29
Cardinal number, 25, 26, 27
  transfinite, 28
Cauchy, A., 155
Cayley, A., 16
Collineation, 139

Complete, 154
Complete quadrangle, 123
Configuration, 121–122
  of Desargues, 124
  of Pappus, 125–126
Congruence invariants, 162
Congruent, 151
Congruently imbeddable, 151
Coordinate set Γ, 58

Dedekind, R., 28
DeMorgan, A., 24
  formulas, 24
Desargues, G., 79
  first property, 79, 80, 81, 84, 85, 87, 88, 89, 90, 92, 93
  second property, 91, 92, 93, 94
  third property, 93, 94, 95, 96
  minor theorem, 125
  theorem, 124, 145–146
Desarguesian plane, 139
Descartes, R., 54
Division ring, 93

Elliptic n-space, axiomatics of, 187
Elliptic plane (model), 185
Euclid, 1, 2, 3, 4, 5, 9, 10, 15, 16, 17, 18, 25, 30, 52, 176
  Elements, 1, 2, 3, 4, 5, 6, 13, 16, 17, 150
Eudoxus, 1

Fano, G., 124
Field, 97

Field (*continued*)
  ordered and continuous, 107–108
  skew, 97
Forder, H. G., 170
Fréchet, M., 150

Gauss, K. F., 9, 11, 12, 13, 14, 176
Geometry, absolute, 53
  affine, 49, 50, 107
  affine analytic over field, 104
  elliptic, 16, 17
  euclidean, 1, 2, 3, 4, 5, 16, 17, 50,
    51
  hyperbolic, 16, 17
  non-euclidean, 12, 15, 16
  parabolic, 16
Gerling, C. L., 14
Group, 84
  affine, 107
  collineation, 139
  projective, 148
  quadruply transitive, 139–140

Hall, M., vi, 64
Heath, T, L., 4, 6
Hessenberg, G., 100
Heyting, A., 37
Hilbert, D., 16, 51, 52, 53
Hyperbolic: line, 177
  plane, 177

Ideal point, 178
Incidence: matrix, 116
  relation, 110

Kant, I., 13
Klein, F., v, 16, 176

Laplace, P. S., 11
Legendre, A. M., 7, 11
Line of slopes, 60
Linear, 157
Lobachewsky, N. I., 9, 14, 15, 176
Logic, 30
Logical connectives, 31
Loop: [Γ, +], 69
  [Γ', ·], 72

Metric space, 150
  subset of, 151

Metrically externally convex, 155
Metrically convex, 155
Moore, E. H., 51

Non-intersectors, 178

Obtuse-angle hypothesis, 10, 13, 15,
  16
One-to-one correspondence, 25, 26,
  27
Oriented line, 163
Origin, 58

Pappian plane, 142
Pappus, 97
  property, 97, 126; affine, 97; spe-
    cial, 98, 99
  theorem, 127; minor, 127
Parallel (fifth) postulate, 2, 4, 52
  substitutes for, 11
Pascal, B., 126
  theorem, 126–127
Pasch, M., 52
Planar ternary ring [Γ, T], 64, 120
  addition in, 68–69
  multiplication in, 70
Plato, 1, 13
Playfair, J., 11, 13, 57
Poincaré, H., 177
Postulates, 2
  for a finite affine geometry, 49–50
  for affine plane, 55
  for elliptic plane, 187
  for euclidean plane geometry, 2–3,
    157, 170
  for hyperbolic plane, 181–182
  for *n*-dimensional euclidean space,
    175
  for projective plane, 110
  for system 7₃, 44
  for three-dimensional euclidean
    geometry (Hilbert's), 51–53
  for two-dimensional spherical
    space, 184–185
Projective plane, 110
  non-Desarguesian, 126
  of order *n*, 115
  over the reals, 111
Projective space of at least three
  dimensions, 144

Postulational system, 38, 39
  categorical, 43
  completeness of, 40, 42, 43
  consistency of, 40, 41
  independence of, 40, 41, 42
  model of, 40, 41
  isomorphism of models, 42, 43
    7₃, 44–49
Principle of duality, 113
Proclus, 5, 6, 11
Proposition, 30, 31
Propositional: calculus, 30
  function, 31

Ray, 167
Rosenthal, A., 51
Right-angle hypothesis, 10
Russel, B., 20, 42
Russell paradox, 20, 21
Russell's aphorism, 39

Saccheri, G., 8, 9, 11, 13, 15, 176
  quadrilateral, 9, 10; summit an-
    gles, 9
Scheffer stroke function, 34
Segment, 169
Set: abstract, 19, 20
  closed, 162
  denumerable, 28
  finite, 28
  infinite, 28
  operations on, 21, 22, 23
Skornyakov, L. A., vi
Slope of line, 60

Spherical geometry (model), 182
Straight line, 160

Taurinus, F. A., 12
Tautology, 34
  deductive theory, 36–37
Ternary: operation T, 61, 120
  system, 120
Thales, 1
Thibaut, B. T., 7, 8
Triangle inequality, 150
Trichotomy theorem, 29
Truth-function, 31
Truth tables, 31, 32, 34

Unit: line, 58
  point, 58

Veblen–Wedderburn: planes, 128
  system, 90
Vector, 73
  addition, 82
  carrier of, 73
  equivalence relation, 73

Whitehead, A. N., 42
Wolfe, H., 51

x-line, 58

y-line, 58

Zermelo, E., 29

A CATALOGUE OF
SELECTED DOVER BOOKS
IN ALL FIELDS OF INTEREST

A CATALOGUE OF SELECTED DOVER
BOOKS IN ALL FIELDS OF INTEREST

CONDITIONED REFLEXES, Ivan P. Pavlov. Full translation of most complete statement of Pavlov's work; cerebral damage, conditioned reflex, experiments with dogs, sleep, similar topics of great importance. 430pp. 5⅜ x 8½. 60614-7 Pa. $4.50

NOTES ON NURSING: WHAT IT IS, AND WHAT IT IS NOT, Florence Nightingale. Outspoken writings by founder of modern nursing. When first published (1860) it played an important role in much needed revolution in nursing. Still stimulating. 140pp. 5⅜ x 8½. 22340-X Pa. $3.00

HARTER'S PICTURE ARCHIVE FOR COLLAGE AND ILLUSTRATION, Jim Harter. Over 300 authentic, rare 19th-century engravings selected by noted collagist for artists, designers, decoupeurs, etc. Machines, people, animals, etc., printed one side of page. 25 scene plates for backgrounds. 6 collages by Harter, Satty, Singer, Evans. Introduction. 192pp. 8⅞ x 11¾. 23659-5 Pa. $5.00

MANUAL OF TRADITIONAL WOOD CARVING, edited by Paul N. Hasluck. Possibly the best book in English on the craft of wood carving. Practical instructions, along with 1,146 working drawings and photographic illustrations. Formerly titled *Cassell's Wood Carving*. 576pp. 6½ x 9¼.
23489-4 Pa. $7.95

THE PRINCIPLES AND PRACTICE OF HAND OR SIMPLE TURNING, John Jacob Holtzapffel. Full coverage of basic lathe techniques—history and development, special apparatus, softwood turning, hardwood turning, metal turning. Many projects—billiard ball, works formed within a sphere, egg cups, ash trays, vases, jardiniers, others—included. 1881 edition. 800 illustrations. 592pp. 6⅛ x 9¼. 23365-0 Clothbd. $15.00

THE JOY OF HANDWEAVING, Osma Tod. Only book you need for hand weaving. Fundamentals, threads, weaves, plus numerous projects for small board-loom, two-harness, tapestry, laid-in, four-harness weaving and more. Over 160 illustrations. 2nd revised edition. 352pp. 6½ x 9¼.
23458-4 Pa. $6.00

THE BOOK OF WOOD CARVING, Charles Marshall Sayers. Still finest book for beginning student in wood sculpture. Noted teacher, craftsman discusses fundamentals, technique; gives 34 designs, over 34 projects for panels, bookends, mirrors, etc. "Absolutely first-rate"—E. J. Tangerman. 33 photos. 118pp. 7¾ x 10⅝. 23654-4 Pa. $3.50

DRAWINGS OF WILLIAM BLAKE, William Blake. 92 plates from Book of Job, *Divine Comedy, Paradise Lost,* visionary heads, mythological figures, Laocoon, etc. Selection, introduction, commentary by Sir Geoffrey Keynes. 178pp. 8⅛ x 11. 22303-5 Pa. $4.00

ENGRAVINGS OF HOGARTH, William Hogarth. 101 of Hogarth's greatest works: *Rake's Progress, Harlot's Progress, Illustrations for Hudibras, Before and After, Beer Street and Gin Lane,* many more. Full commentary. 256pp. 11 x 13¾. 22479-1 Pa. $12.95

DAUMIER: 120 GREAT LITHOGRAPHS, Honore Daumier. Wide-ranging collection of lithographs by the greatest caricaturist of the 19th century. Concentrates on eternally popular series on lawyers, on married life, on liberated women, etc. Selection, introduction, and notes on plates by Charles F. Ramus. Total of 158pp. 9⅜ x 12¼. 23512-2 Pa. $6.00

DRAWINGS OF MUCHA, Alphonse Maria Mucha. Work reveals drafts-man of highest caliber: studies for famous posters and paintings, render-ings for book illustrations and ads, etc. 70 works, 9 in color; including 6 items not drawings. Introduction. List of illustrations. 72pp. 9⅜ x 12¼. (Available in U.S. only) 23672-2 Pa. $4.00

GIOVANNI BATTISTA PIRANESI: DRAWINGS IN THE PIERPONT MORGAN LIBRARY, Giovanni Battista Piranesi. For first time ever all of Morgan Library's collection, world's largest. 167 illustrations of rare Piranesi drawings—archeological, architectural, decorative and visionary. Essay, detailed list of drawings, chronology, captions. Edited by Felice Stampfle. 144pp. 9⅜ x 12¼. 23714-1 Pa. $7.50

NEW YORK ETCHINGS (1905-1949), John Sloan. All of important American artist's N.Y. life etchings. 67 works include some of his best art; also lively historical record—Greenwich Village, tenement scenes. Edited by Sloan's widow. Introduction and captions. 79pp. 8⅜ x 11¼. 23651-X Pa. $4.00

CHINESE PAINTING AND CALLIGRAPHY: A PICTORIAL SURVEY, Wan-go Weng. 69 fine examples from John M. Crawford's matchless private collection: landscapes, birds, flowers, human figures, etc., plus calligraphy. Every basic form included: hanging scrolls, handscrolls, album leaves, fans, etc. 109 illustrations. Introduction. Captions. 192pp. 8⅞ x 11¾. 23707-9 Pa. $7.95

DRAWINGS OF REMBRANDT, edited by Seymour Slive. Updated Lipp-mann, Hofstede de Groot edition, with definitive scholarly apparatus. All portraits, biblical sketches, landscapes, nudes, Oriental figures, classical studies, together with selection of work by followers. 550 illustrations. Total of 630pp. 9⅛ x 12¼. 21485-0, 21486-9 Pa., Two-vol. set $15.00

THE DISASTERS OF WAR, Francisco Goya. 83 etchings record horrors of Napoleonic wars in Spain and war in general. Reprint of 1st edition, plus 3 additional plates. Introduction by Philip Hofer. 97pp. 9⅜ x 8¼. 21872-4 Pa. $4.00

THE AMERICAN SENATOR, Anthony Trollope. Little known, long unavailable Trollope novel on a grand scale. Here are humorous comment on American vs. English culture, and stunning portrayal of a heroine/villainess. Superb evocation of Victorian village life. 561pp. 5⅜ x 8½.
23801-6 Pa. $6.00

WAS IT MURDER? James Hilton. The author of *Lost Horizon* and *Goodbye, Mr. Chips* wrote one detective novel (under a pen-name) which was quickly forgotten and virtually lost, even at the height of Hilton's fame. This edition brings it back—a finely crafted public school puzzle resplendent with Hilton's stylish atmosphere. A thoroughly English thriller by the creator of Shangri-la. 252pp. 5⅜ x 8. (Available in U.S. only)
23774-5 Pa. $3.00

CENTRAL PARK: A PHOTOGRAPHIC GUIDE, Victor Laredo and Henry Hope Reed. 121 superb photographs show dramatic views of Central Park: Bethesda Fountain, Cleopatra's Needle, Sheep Meadow, the Blockhouse, plus people engaged in many park activities: ice skating, bike riding, etc. Captions by former Curator of Central Park, Henry Hope Reed, provide historical view, changes, etc. Also photos of N.Y. landmarks on park's periphery. 96pp. 8½ x 11. 23750-8 Pa. $4.50

NANTUCKET IN THE NINETEENTH CENTURY, Clay Lancaster. 180 rare photographs, stereographs, maps, drawings and floor plans recreate unique American island society. Authentic scenes of shipwreck, lighthouses, streets, homes are arranged in geographic sequence to provide walking-tour guide to old Nantucket existing today. Introduction, captions. 160pp. 8⅞ x 11¾. 23747-8 Pa. $6.95

STONE AND MAN: A PHOTOGRAPHIC EXPLORATION, Andreas Feininger. 106 photographs by *Life* photographer Feininger portray man's deep passion for stone through the ages. Stonehenge-like megaliths, fortified towns, sculpted marble and crumbling tenements show textures, beauties, fascination. 128pp. 9¼ x 10¾. 23756-7 Pa. $5.95

CIRCLES, A MATHEMATICAL VIEW, D. Pedoe. Fundamental aspects of college geometry, non-Euclidean geometry, and other branches of mathematics: representing circle by point. Poincare model, isoperimetric property, etc. Stimulating recreational reading. 66 figures. 96pp. 5⅝ x 8¼.
63698-4 Pa. $2.75

THE DISCOVERY OF NEPTUNE, Morton Grosser. Dramatic scientific history of the investigations leading up to the actual discovery of the eighth planet of our solar system. Lucid, well-researched book by well-known historian of science. 172pp. 5⅜ x 8½. 23726-5 Pa. $3.50

THE DEVIL'S DICTIONARY. Ambrose Bierce. Barbed, bitter, brilliant witticisms in the form of a dictionary. Best, most ferocious satire America has produced. 145pp. 5⅜ x 8½. 20487-1 Pa. $2.25

HISTORY OF BACTERIOLOGY, William Bulloch. The only comprehensive history of bacteriology from the beginnings through the 19th century. Special emphasis is given to biography-Leeuwenhoek, etc. Brief accounts of 350 bacteriologists form a separate section. No clearer, fuller study, suitable to scientists and general readers, has yet been written. 52 illustrations. 448pp. 5⅝ x 8¼.   23761-3 Pa. $6.50

THE COMPLETE NONSENSE OF EDWARD LEAR, Edward Lear. All nonsense limericks, zany alphabets, Owl and Pussycat, songs, nonsense botany, etc., illustrated by Lear. Total of 321pp. 5⅜ x 8½. (Available in U.S. only)   20167-8 Pa. $3.95

INGENIOUS MATHEMATICAL PROBLEMS AND METHODS, Louis A. Graham. Sophisticated material from Graham *Dial*, applied and pure; stresses solution methods. Logic, number theory, networks, inversions, etc. 237pp. 5⅜ x 8½.   20545-2 Pa. $4.50

BEST MATHEMATICAL PUZZLES OF SAM LOYD, edited by Martin Gardner. Bizarre, original, whimsical puzzles by America's greatest puzzler. From fabulously rare *Cyclopedia*, including famous 14-15 puzzles, the Horse of a Different Color, 115 more. Elementary math. 150 illustrations. 167pp. 5⅜ x 8½.   20498-7 Pa. $2.75

THE BASIS OF COMBINATION IN CHESS, J. du Mont. Easy-to-follow, instructive book on elements of combination play, with chapters on each piece and every powerful combination team—two knights, bishop and knight, rook and bishop, etc. 250 diagrams. 218pp. 5⅜ x 8½. (Available in U.S. only)   23644-7 Pa. $3.50

MODERN CHESS STRATEGY, Ludek Pachman. The use of the queen, the active king, exchanges, pawn play, the center, weak squares, etc. Section on rook alone worth price of the book. Stress on the moderns. Often considered the most important book on strategy. 314pp. 5⅜ x 8½.
                                                    20290-9 Pa. $4.50

LASKER'S MANUAL OF CHESS, Dr. Emanuel Lasker. Great world champion offers very thorough coverage of all aspects of chess. Combinations, position play, openings, end game, aesthetics of chess, philosophy of struggle, much more. Filled with analyzed games. 390pp. 5⅜ x 8½.
                                                    20640-8 Pa. $5.00

500 MASTER GAMES OF CHESS, S. Tartakower, J. du Mont. Vast collection of great chess games from 1798-1938, with much material nowhere else readily available. Fully annotated, arranged by opening for easier study. 664pp. 5⅜ x 8½.   23208-5 Pa. $7.50

A GUIDE TO CHESS ENDINGS, Dr. Max Euwe, David Hooper. One of the finest modern works on chess endings. Thorough analysis of the most frequently encountered endings by former world champion. 331 examples, each with diagram. 248pp. 5⅜ x 8½.   23332-4 Pa. $3.75

SECOND PIATIGORSKY CUP, edited by Isaac Kashdan. One of the greatest tournament books ever produced in the English language. All 90 games of the 1966 tournament, annotated by players, most annotated by both players. Features Petrosian, Spassky, Fischer, Larsen, six others. 228pp. 5⅜ x 8½. 23572-6 Pa. $3.50

ENCYCLOPEDIA OF CARD TRICKS, revised and edited by Jean Hugard. How to perform over 600 card tricks, devised by the world's greatest magicians: impromptus, spelling tricks, key cards, using special packs, much, much more. Additional chapter on card technique. 66 illustrations. 402pp. 5⅜ x 8½. (Available in U.S. only) 21252-1 Pa. $4.95

MAGIC: STAGE ILLUSIONS, SPECIAL EFFECTS AND TRICK PHO-TOGRAPHY, Albert A. Hopkins, Henry R. Evans. One of the great classics; fullest, most authorative explanation of vanishing lady, levitations, scores of other great stage effects. Also small magic, automata, stunts. 446 illustrations. 556pp. 5⅜ x 8½. 23344-8 Pa. $6.95

THE SECRETS OF HOUDINI, J. C. Cannell. Classic study of Houdini's incredible magic, exposing closely-kept professional secrets and revealing, in general terms, the whole art of stage magic. 67 illustrations. 279pp. 5⅜ x 8½. 22913-0 Pa. $4.00

HOFFMANN'S MODERN MAGIC, Professor Hoffmann. One of the best, and best-known, magicians' manuals of the past century. Hundreds of tricks from card tricks and simple sleight of hand to elaborate illusions involving construction of complicated machinery. 332 illustrations. 563pp. 5⅜ x 8½. 23623-4 Pa. $6.00

MADAME PRUNIER'S FISH COOKERY BOOK, Mme. S. B. Prunier. More than 1000 recipes from world famous Prunier's of Paris and London, specially adapted here for American kitchen. Grilled tournedos with anchovy butter, Lobster a la Bordelaise, Prunier's prized desserts, more. Glossary. 340pp. 5⅜ x 8½. (Available in U.S. only) 22679-4 Pa. $3.00

FRENCH COUNTRY COOKING FOR AMERICANS, Louis Diat. 500 easy-to-make, authentic provincial recipes compiled by former head chef at New York's Fitz-Carlton Hotel: onion soup, lamb stew, potato pie, more. 309pp. 5⅜ x 8½. 23665-X Pa. $3.95

SAUCES, FRENCH AND FAMOUS, Louis Diat. Complete book gives over 200 specific recipes: bechamel, Bordelaise, hollandaise, Cumberland, apricot, etc. Author was one of this century's finest chefs, originator of vichyssoise and many other dishes. Index. 156pp. 5⅜ x 8. 23663-3 Pa. $2.75

TOLL HOUSE TRIED AND TRUE RECIPES, Ruth Graves Wakefield. Authentic recipes from the famous Mass. restaurant: popovers, veal and ham loaf, Toll House baked beans, chocolate cake crumb pudding, much more. Many helpful hints. Nearly 700 recipes. Index. 376pp. 5⅜ x 8½. 23560-2 Pa. $4.50

YUCATAN BEFORE AND AFTER THE CONQUEST, Diego de Landa. First English translation of basic book in Maya studies, the only significant account of Yucatan written in the early post-Conquest era. Translated by distinguished Maya scholar William Gates. Appendices, introduction, 4 maps and over 120 illustrations added by translator. 162pp. 5⅜ x 8½.
23622-6 Pa. $3.00

THE MALAY ARCHIPELAGO, Alfred R. Wallace. Spirited travel account by one of founders of modern biology. Touches on zoology, botany, ethnography, geography, and geology. 62 illustrations, maps. 515pp. 5⅜ x 8½.
20187-2 Pa. $6.95

THE DISCOVERY OF THE TOMB OF TUTANKHAMEN, Howard Carter, A. C. Mace. Accompany Carter in the thrill of discovery, as ruined passage suddenly reveals unique, untouched, fabulously rich tomb. Fascinating account, with 106 illustrations. New introduction by J. M. White. Total of 382pp. 5⅜ x 8½. (Available in U.S. only)   23500-9 Pa. $4.00

THE WORLD'S GREATEST SPEECHES, edited by Lewis Copeland and Lawrence W. Lamm. Vast collection of 278 speeches from Greeks up to present. Powerful and effective models; unique look at history. Revised to 1970. Indices. 842pp. 5⅜ x 8½.   20468-5 Pa. $8.95

THE 100 GREATEST ADVERTISEMENTS, Julian Watkins. The priceless ingredient; His master's voice; 99 44/100% pure; over 100 others. How they were written, their impact, etc. Remarkable record. 130 illustrations. 233pp. 7⅞ x 10 3/5.   20540-1 Pa. $5.95

CRUICKSHANK PRINTS FOR HAND COLORING, George Cruickshank. 18 illustrations, one side of a page, on fine-quality paper suitable for watercolors. Caricatures of people in society (c. 1820) full of trenchant wit. Very large format. 32pp. 11 x 16.   23684-6 Pa. $5.00

THIRTY-TWO COLOR POSTCARDS OF TWENTIETH-CENTURY AMERICAN ART, Whitney Museum of American Art. Reproduced in full color in postcard form are 31 art works and one shot of the museum. Calder, Hopper, Rauschenberg, others. Detachable. 16pp. 8¼ x 11.
23629-3 Pa. $3.00

MUSIC OF THE SPHERES: THE MATERIAL UNIVERSE FROM ATOM TO QUASAR SIMPLY EXPLAINED, Guy Murchie. Planets, stars, geology, atoms, radiation, relativity, quantum theory, light, antimatter, similar topics. 319 figures. 664pp. 5⅜ x 8½.
21809-0, 21810-4 Pa., Two-vol. set $11.00

EINSTEIN'S THEORY OF RELATIVITY, Max Born. Finest semi-technical account; covers Einstein, Lorentz, Minkowski, and others, with much detail, much explanation of ideas and math not readily available elsewhere on this level. For student, non-specialist. 376pp. 5⅜ x 8½.
60769-0 Pa. $4.50

THE EARLY WORK OF AUBREY BEARDSLEY, Aubrey Beardsley. 157 plates, 2 in color: *Manon Lescaut, Madame Bovary, Morte Darthur, Salome,* other. Introduction by H. Marillier. 182pp. 8⅛ x 11. 21816-3 Pa. $4.50

THE LATER WORK OF AUBREY BEARDSLEY, Aubrey Beardsley. Exotic masterpieces of full maturity: *Venus and Tannhauser, Lysistrata, Rape of the Lock, Volpone,* Savoy material, etc. 174 plates, 2 in color. 186pp. 8⅛ x 11. 21817-1 Pa. $5.95

THOMAS NAST'S CHRISTMAS DRAWINGS, Thomas Nast. Almost all Christmas drawings by creator of image of Santa Claus as we know it, and one of America's foremost illustrators and political cartoonists. 66 illustrations. 3 illustrations in color on covers. 96pp. 8⅜ x 11¼. 23660-9 Pa. $3.50

THE DORÉ ILLUSTRATIONS FOR DANTE'S DIVINE COMEDY, Gustave Doré. All 135 plates from Inferno, Purgatory, Paradise; fantastic tortures, infernal landscapes, celestial wonders. Each plate with appropriate (translated) verses. 141pp. 9 x 12. 23231-X Pa. $4.50

DORÉ'S ILLUSTRATIONS FOR RABELAIS, Gustave Doré. 252 striking illustrations of *Gargantua and Pantagruel* books by foremost 19th-century illustrator. Including 60 plates, 192 delightful smaller illustrations. 153pp. 9 x 12. 23656-0 Pa. $5.00

LONDON: A PILGRIMAGE, Gustave Doré, Blanchard Jerrold. Squalor, riches, misery, beauty of mid-Victorian metropolis; 55 wonderful plates, 125 other illustrations, full social, cultural text by Jerrold. 191pp. of text. 9⅜ x 12¼. 22306-X Pa. $7.00

THE RIME OF THE ANCIENT MARINER, Gustave Doré, S. T. Coleridge. Dore's finest work, 34 plates capture moods, subtleties of poem. Full text. Introduction by Millicent Rose. 77pp. 9¼ x 12. 22305-1 Pa. $3.50

THE DORE BIBLE ILLUSTRATIONS, Gustave Doré. All wonderful, detailed plates: Adam and Eve, Flood, Babylon, Life of Jesus, etc. Brief King James text with each plate. Introduction by Millicent Rose. 241 plates. 241pp. 9 x 12. 23004-X Pa. $6.00

THE COMPLETE ENGRAVINGS, ETCHINGS AND DRYPOINTS OF ALBRECHT DURER. "Knight, Death and Devil"; "Melencolia," and more—all Dürer's known works in all three media, including 6 works formerly attributed to him. 120 plates. 235pp. 8⅜ x 11¼. 22851-7 Pa. $6.50

MECHANICK EXERCISES ON THE WHOLE ART OF PRINTING, Joseph Moxon. First complete book (1683-4) ever written about typography, a compendium of everything known about printing at the latter part of 17th century. Reprint of 2nd (1962) Oxford Univ. Press edition. 74 illustrations. Total of 550pp. 6⅛ x 9¼. 23617-X Pa. $7.95

THE COMPLETE WOODCUTS OF ALBRECHT DURER, edited by Dr. W. Kurth. 346 in all: "Old Testament," "St. Jerome," "Passion," "Life of Virgin," Apocalypse," many others. Introduction by Campbell Dodgson. 285pp. 8½ x 12¼. 21097-9 Pa. $7.50

DRAWINGS OF ALBRECHT DURER, edited by Heinrich Wolfflin. 81 plates show development from youth to full style. Many favorites; many new. Introduction by Alfred Werner. 96pp. 8⅛ x 11. 22352-3 Pa. $5.00

THE HUMAN FIGURE, Albrecht Dürer. Experiments in various techniques—stereometric, progressive proportional, and others. Also life studies that rank among finest ever done. Complete reprinting of Dresden Sketchbook. 170 plates. 355pp. 8⅜ x 11¼. 21042-1 Pa. $7.95

OF THE JUST SHAPING OF LETTERS, Albrecht Dürer. Renaissance artist explains design of Roman majuscules by geometry, also Gothic lower and capitals. Grolier Club edition. 43pp. 7⅞ x 10¾ 21306-4 Pa. $3.00

TEN BOOKS ON ARCHITECTURE, Vitruvius. The most important book ever written on architecture. Early Roman aesthetics, technology, classical orders, site selection, all other aspects. Stands behind everything since. Morgan translation. 331pp. 5⅜ x 8½. 20645-9 Pa. $4.50

THE FOUR BOOKS OF ARCHITECTURE, Andrea Palladio. 16th-century classic responsible for Palladian movement and style. Covers classical architectural remains, Renaissance revivals, classical orders, etc. 1738 Ware English edition. Introduction by A. Placzek. 216 plates. 110pp. of text. 9½ x 12¾. 21308-0 Pa. $10.00

HORIZONS, Norman Bel Geddes. Great industrialist stage designer, "father of streamlining," on application of aesthetics to transportation, amusement, architecture, etc. 1932 prophetic account; function, theory, specific projects. 222 illustrations. 312pp. 7⅞ x 10¾. 23514-9 Pa. $6.95

FRANK LLOYD WRIGHT'S FALLINGWATER, Donald Hoffmann. Full, illustrated story of conception and building of Wright's masterwork at Bear Run, Pa. 100 photographs of site, construction, and details of completed structure. 112pp. 9¼ x 10. 23671-4 Pa. $5.50

THE ELEMENTS OF DRAWING, John Ruskin. Timeless classic by great Viltorian; starts with basic ideas, works through more difficult. Many practical exercises. 48 illustrations. Introduction by Lawrence Campbell. 228pp. 5⅜ x 8½. 22730-8 Pa. $3.75

GIST OF ART, John Sloan. Greatest modern American teacher, Art Students League, offers innumerable hints, instructions, guided comments to help you in painting. Not a formal course. 46 illustrations. Introduction by Helen Sloan. 200pp. 5⅜ x 8½. 23435-5 Pa. $4.00

THE ANATOMY OF THE HORSE, George Stubbs. Often considered the great masterpiece of animal anatomy. Full reproduction of 1766 edition, plus prospectus; original text and modernized text. 36 plates. Introduction by Eleanor Garvey. 121pp. 11 x 14¾.   23402-9 Pa. $6.00

BRIDGMAN'S LIFE DRAWING, George B. Bridgman. More than 500 illustrative drawings and text teach you to abstract the body into its major masses, use light and shade, proportion; as well as specific areas of anatomy, of which Bridgman is master. 192pp. 6½ x 9¼. (Available in U.S. only)
22710-3 Pa. $3.50

ART NOUVEAU DESIGNS IN COLOR, Alphonse Mucha, Maurice Verneuil, Georges Auriol. Full-color reproduction of *Combinaisons ornementales* (c. 1900) by Art Nouveau masters. Floral, animal, geometric, interlacings, swashes—borders, frames, spots—all incredibly beautiful. 60 plates, hundreds of designs. 9⅜ x 8-1/16.   22885-1 Pa. $4.00

FULL-COLOR FLORAL DESIGNS IN THE ART NOUVEAU STYLE, E. A. Seguy. 166 motifs, on 40 plates, from *Les fleurs et leurs applications decoratives* (1902): borders, circular designs, repeats, allovers, "spots." All in authentic Art Nouveau colors. 48pp. 9⅜ x 12¼.
23439-8 Pa. $5.00

A DIDEROT PICTORIAL ENCYCLOPEDIA OF TRADES AND IN-DUSTRY, edited by Charles C. Gillispie. 485 most interesting plates from the great French Encyclopedia of the 18th century show hundreds of working figures, artifacts, process, land and cityscapes; glassmaking, paper-making, metal extraction, construction, weaving, making furniture, clothing, wigs, dozens of other activities. Plates fully explained. 920pp. 9 x 12.
22284-5, 22285-3 Clothbd., Two-vol. set $40.00

HANDBOOK OF EARLY ADVERTISING ART, Clarence P. Hornung. Largest collection of copyright-free early and antique advertising art ever compiled. Over 6,000 illustrations, from Franklin's time to the 1890's for special effects, novelty. Valuable source, almost inexhaustible.
Pictorial Volume. Agriculture, the zodiac, animals, autos, birds, Christmas, fire engines, flowers, trees, musical instruments, ships, games and sports, much more. Arranged by subject matter and use. 237 plates. 288pp. 9 x 12.
20122-8 Clothbd. $14.50

Typographical Volume. Roman and Gothic faces ranging from 10 point to 300 point, "Barnum," German and Old English faces, script, logotypes, scrolls and flourishes, 1115 ornamental initials, 67 complete alphabets, more. 310 plates. 320pp. 9 x 12.   20123-6 Clothbd. $15.00

CALLIGRAPHY (CALLIGRAPHIA LATINA), J. G. Schwandner. High point of 18th-century ornamental calligraphy. Very ornate initials, scrolls, borders, cherubs, birds, lettered examples. 172pp. 9 x 13.
20475-8 Pa. $7.00

HOLLYWOOD GLAMOUR PORTRAITS, edited by John Kobal. 145 photos capture the stars from 1926-49, the high point in portrait photography. Gable, Harlow, Bogart, Bacall, Hedy Lamarr, Marlene Dietrich, Robert Montgomery, Marlon Brando, Veronica Lake; 94 stars in all. Full background on photographers, technical aspects, much more. Total of 160pp. 8⅜ x 11¼. 23352-9 Pa. $6.00

THE NEW YORK STAGE: FAMOUS PRODUCTIONS IN PHOTO-GRAPHS, edited by Stanley Appelbaum. 148 photographs from Museum of City of New York show 142 plays, 1883-1939. *Peter Pan, The Front Page, Dead End, Our Town,* O'Neill, hundreds of actors and actresses, etc. Full indexes. 154pp. 9½ x 10. 23241-7 Pa. $6.00

DIALOGUES CONCERNING TWO NEW SCIENCES, Galileo Galilei. Encompassing 30 years of experiment and thought, these dialogues deal with geometric demonstrations of fracture of solid bodies, cohesion, leverage, speed of light and sound, pendulums, falling bodies, accelerated motion, etc. 300pp. 5⅜ x 8½. 60099-8 Pa. $4.00

THE GREAT OPERA STARS IN HISTORIC PHOTOGRAPHS, edited by James Camner. 343 portraits from the 1850s to the 1940s: Tamburini, Mario, Caliapin, Jeritza, Melchior, Melba, Patti, Pinza, Schipa, Caruso, Farrar, Steber, Gobbi, and many more—270 performers in all. Index. 199pp. 8⅜ x 11¼. 23575-0 Pa. $7.50

J. S. BACH, Albert Schweitzer. Great full-length study of Bach, life, background to music, music, by foremost modern scholar. Ernest Newman translation. 650 musical examples. Total of 928pp. 5⅜ x 8½. (Available in U.S. only) 21631-4, 21632-2 Pa., Two-vol. set $11.00

COMPLETE PIANO SONATAS, Ludwig van Beethoven. All sonatas in the fine Schenker edition, with fingering, analytical material. One of best modern editions. Total of 615pp. 9 x 12. (Available in U.S. only) 23134-8, 23135-6 Pa., Two-vol. set $15.50

KEYBOARD MUSIC, J. S. Bach. Bach-Gesellschaft edition. For harpsichord, piano, other keyboard instruments. English Suites, French Suites, Six Partitas, Goldberg Variations, Two-Part Inventions, Three-Part Sinfonias. 312pp. 8⅛ x 11. (Available in U.S. only) 22360-4 Pa. $6.95

FOUR SYMPHONIES IN FULL SCORE, Franz Schubert. Schubert's four most popular symphonies: No. 4 in C Minor ("Tragic"); No. 5 in B-flat Major; No. 8 in B Minor ("Unfinished"); No. 9 in C Major ("Great"). Breitkopf & Hartel edition. Study score. 261pp. 9⅜ x 12¼.
23681-1 Pa. $6.50

THE AUTHENTIC GILBERT & SULLIVAN SONGBOOK, W. S. Gilbert, A. S. Sullivan. Largest selection available; 92 songs, uncut, original keys, in piano rendering approved by Sullivan. Favorites and lesser-known fine numbers. Edited with plot synopses by James Spero. 3 illustrations. 399pp. 9 x 12. 23482-7 Pa. $9.95

A MAYA GRAMMAR, Alfred M. Tozzer. Practical, useful English-language grammar by the Harvard anthropologist who was one of the three greatest American scholars in the area of Maya culture. Phonetics, grammatical processes, syntax, more. 301pp. 5⅜ x 8½. 23465-7 Pa. $4.00

THE JOURNAL OF HENRY D. THOREAU, edited by Bradford Torrey, F. H. Allen. Complete reprinting of 14 volumes, 1837-61, over two million words; the sourcebooks for *Walden*, etc. Definitive. All original sketches, plus 75 photographs. Introduction by Walter Harding. Total of 1804pp. 8½ x 12¼. 20312-3, 20313-1 Clothbd., Two-vol. set $70.00

CLASSIC GHOST STORIES, Charles Dickens and others. 18 wonderful stories you've wanted to reread: "The Monkey's Paw," "The House and the Brain," "The Upper Berth," "The Signalman," "Dracula's Guest," "The Tapestried Chamber," etc. Dickens, Scott, Mary Shelley, Stoker, etc. 330pp. 5⅜ x 8½. 20735-8 Pa. $4.50

SEVEN SCIENCE FICTION NOVELS, H. G. Wells. Full novels. *First Men in the Moon, Island of Dr. Moreau, War of the Worlds, Food of the Gods, Invisible Man, Time Machine, In the Days of the Comet.* A basic science-fiction library. 1015pp. 5⅜ x 8½. (Available in U.S. only) 20264-X Clothbd. $8.95

ARMADALE, Wilkie Collins. Third great mystery novel by the author of *The Woman in White* and *The Moonstone.* Ingeniously plotted narrative shows an exceptional command of character, incident and mood. Original magazine version with 40 illustrations. 597pp. 5⅜ x 8½. 23429-0 Pa. $6.00

MASTERS OF MYSTERY, H. Douglas Thomson. The first book in English (1931) devoted to history and aesthetics of detective story. Poe, Doyle, LeFanu, Dickens, many others, up to 1930. New introduction and notes by E. F. Bleiler. 288pp. 5⅜ x 8½. (Available in U.S. only) 23606-4 Pa. $4.00

FLATLAND, E. A. Abbott. Science-fiction classic explores life of 2-D being in 3-D world. Read also as introduction to thought about hyperspace. Introduction by Banesh Hoffmann. 16 illustrations. 103pp. 5⅜ x 8½. 20001-9 Pa. $2.00

THREE SUPERNATURAL NOVELS OF THE VICTORIAN PERIOD, edited, with an introduction, by E. F. Bleiler. Reprinted complete and unabridged, three great classics of the supernatural: *The Haunted Hotel* by Wilkie Collins, *The Haunted House at Latchford* by Mrs. J. H. Riddell, and *The Lost Stradivarius* by J. Meade Falkner. 325pp. 5⅜ x 8½. 22571-2 Pa. $4.00

AYESHA: THE RETURN OF "SHE," H. Rider Haggard. Virtuoso sequel featuring the great mythic creation, Ayesha, in an adventure that is fully as good as the first book, *She.* Original magazine version, with 47 original illustrations by Maurice Greiffenhagen. 189pp. 6½ x 9¼. 23649-8 Pa. $3.50

HOUSEHOLD STORIES BY THE BROTHERS GRIMM. All the great Grimm stories: "Rumpelstiltskin," "Snow White," "Hansel and Gretel," etc., with 114 illustrations by Walter Crane. 269pp. 5⅜ x 8½.
21080-4 Pa. $3.50

SLEEPING BEAUTY, illustrated by Arthur Rackham. Perhaps the fullest, most delightful version ever, told by C. S. Evans. Rackham's best work. 49 illustrations. 110pp. 7⅞ x 10¾.
22756-1 Pa. $2.50

AMERICAN FAIRY TALES, L. Frank Baum. Young cowboy lassoes Father Time; dummy in Mr. Floman's department store window comes to life; and 10 other fairy tales. 41 illustrations by N. P. Hall, Harry Kennedy, Ike Morgan, and Ralph Gardner. 209pp. 5⅜ x 8½.
23643-9 Pa. $3.00

THE WONDERFUL WIZARD OF OZ, L. Frank Baum. Facsimile in full color of America's finest children's classic. Introduction by Martin Gardner. 143 illustrations by W. W. Denslow. 267pp. 5⅜ x 8½.
20691-2 Pa. $3.50

THE TALE OF PETER RABBIT, Beatrix Potter. The inimitable Peter's terrifying adventure in Mr. McGregor's garden, with all 27 wonderful, full-color Potter illustrations. 55pp. 4¼ x 5½. (Available in U.S. only)
22827-4 Pa. $1.25

THE STORY OF KING ARTHUR AND HIS KNIGHTS, Howard Pyle. Finest children's version of life of King Arthur. 48 illustrations by Pyle. 131pp. 6⅛ x 9¼.
21445-1 Pa. $4.95

CARUSO'S CARICATURES, Enrico Caruso. Great tenor's remarkable caricatures of self, fellow musicians, composers, others. Toscanini, Puccini, Farrar, etc. Impish, cutting, insightful. 473 illustrations. Preface by M. Sisca. 217pp. 8⅜ x 11¼.
23528-9 Pa. $6.95

PERSONAL NARRATIVE OF A PILGRIMAGE TO ALMADINAH AND MECCAH, Richard Burton. Great travel classic by remarkably colorful personality. Burton, disguised as a Moroccan, visited sacred shrines of Islam, narrowly escaping death. Wonderful observations of Islamic life, customs, personalities. 47 illustrations. Total of 959pp. 5⅜ x 8½.
21217-3, 21218-1 Pa., Two-vol. set $12.00

INCIDENTS OF TRAVEL IN YUCATAN, John L. Stephens. Classic (1843) exploration of jungles of Yucatan, looking for evidences of Maya civilization. Travel adventures, Mexican and Indian culture, etc. Total of 669pp. 5⅜ x 8½.
20926-1, 20927-X Pa., Two-vol. set $7.90

AMERICAN LITERARY AUTOGRAPHS FROM WASHINGTON IRVING TO HENRY JAMES, Herbert Cahoon, et al. Letters, poems, manuscripts of Hawthorne, Thoreau, Twain, Alcott, Whitman, 67 other prominent American authors. Reproductions, full transcripts and commentary. Plus checklist of all American Literary Autographs in The Pierpont Morgan Library. Printed on exceptionally high-quality paper. 136 illustrations. 212pp. 9⅛ x 12¼.
23548-3 Pa. $12.50

PRINCIPLES OF ORCHESTRATION, Nikolay Rimsky-Korsakov. Great classical orchestrator provides fundamentals of tonal resonance, progression of parts, voice and orchestra, tutti effects, much else in major document. 330pp. of musical excerpts. 489pp. 6½ x 9¼. 21266-1 Pa. $7.50

TRISTAN UND ISOLDE, Richard Wagner. Full orchestral score with complete instrumentation. Do not confuse with piano reduction. Commentary by Felix Mottl, great Wagnerian conductor and scholar. Study score. 655pp. 8⅛ x 11. 22915-7 Pa. $13.95

REQUIEM IN FULL SCORE, Giuseppe Verdi. Immensely popular with choral groups and music lovers. Republication of edition published by C. F. Peters, Leipzig, n. d. German frontmaker in English translation. Glossary. Text in Latin. Study score. 204pp. 9⅜ x 12¼.
23682-X Pa. $6.00

COMPLETE CHAMBER MUSIC FOR STRINGS, Felix Mendelssohn. All of Mendelssohn's chamber music: Octet, 2 Quintets, 6 Quartets, and Four Pieces for String Quartet. (Nothing with piano is included). Complete works edition (1874-7). Study score. 283 pp. 9⅜ x 12¼.
23679-X Pa. $7.50

POPULAR SONGS OF NINETEENTH-CENTURY AMERICA, edited by Richard Jackson. 64 most important songs: "Old Oaken Bucket," "Arkansas Traveler," "Yellow Rose of Texas," etc. Authentic original sheet music, full introduction and commentaries. 290pp. 9 x 12. 23270-0 Pa. $7.95

COLLECTED PIANO WORKS, Scott Joplin. Edited by Vera Brodsky Lawrence. Practically all of Joplin's piano works—rags, two-steps, marches, waltzes, etc., 51 works in all. Extensive introduction by Rudi Blesh. Total of 345pp. 9 x 12. 23106-2 Pa. $14.95

BASIC PRINCIPLES OF CLASSICAL BALLET, Agrippina Vaganova. Great Russian theoretician, teacher explains methods for teaching classical ballet; incorporates best from French, Italian, Russian schools. 118 illustrations. 175pp. 5⅜ x 8½. 22036-2 Pa. $2.50

CHINESE CHARACTERS, L. Wieger. Rich analysis of 2300 characters according to traditional systems into primitives. Historical-semantic analysis to phonetics (Classical Mandarin) and radicals. 820pp. 6⅛ x 9¼.
21321-8 Pa. $10.00

EGYPTIAN LANGUAGE: EASY LESSONS IN EGYPTIAN HIERO-GLYPHICS, E. A. Wallis Budge. Foremost Egyptologist offers Egyptian grammar, explanation of hieroglyphics, many reading texts, dictionary of symbols. 246pp. 5 x 7½. (Available in U.S. only)
21394-3 Clothbd. $7.50

AN ETYMOLOGICAL DICTIONARY OF MODERN ENGLISH, Ernest Weekley. Richest, fullest work, by foremost British lexicographer. Detailed word histories. Inexhaustible. Do not confuse this with *Concise Etymological Dictionary,* which is abridged. Total of 856pp. 6½ x 9¼.
21873-2, 21874-0 Pa., Two-vol. set $12.00

GEOMETRY, RELATIVITY AND THE FOURTH DIMENSION, Rudolf Rucker. Exposition of fourth dimension, means of visualization, concepts of relativity as Flatland characters continue adventures. Popular, easily followed yet accurate, profound. 141 illustrations. 133pp. 5⅜ x 8½.
23400-2 Pa. $2.75

THE ORIGIN OF LIFE, A. I. Oparin. Modern classic in biochemistry, the first rigorous examination of possible evolution of life from nitrocarbon compounds. Non-technical, easily followed. Total of 295pp. 5⅜ x 8½.
60213-3 Pa. $4.00

PLANETS, STARS AND GALAXIES, A. E. Fanning. Comprehensive introductory survey: the sun, solar system, stars, galaxies, universe, cosmology; quasars, radio stars, etc. 24pp. of photographs. 189pp. 5⅜ x 8½. (Available in U.S. only)
21680-2 Pa. $3.75

THE THIRTEEN BOOKS OF EUCLID'S ELEMENTS, translated with introduction and commentary by Sir Thomas L. Heath. Definitive edition. Textual and linguistic notes, mathematical analysis, 2500 years of critical commentary. Do not confuse with abridged school editions. Total of 1414pp. 5⅜ x 8½.
60088-2, 60089-0, 60090-4 Pa., Three-vol. set $18.50

Prices subject to change without notice.

Available at your book dealer or write for free catalogue to Dept. GI, Dover Publications, Inc., 180 Varick St., N.Y., N.Y. 10014. Dover publishes more than 175 books each year on science, elementary and advanced mathematics, biology, music, art, literary history, social sciences and other areas.